CONSCIOUSNESS

By Joseph Magrie

Consciousness

Delta Δ Books

ISBN: 978-0-9987552-4-3

For

Tina, Heather, and Quinn

Table of Contents

Table of Contents

Foreword

What is consciousness? How did it arise? Why did it evolve? These are among the most profound questions ever asked. More than three thousand years ago consciousness was described in the Rig Veda as being a fundamental property of human life. Since then, numerous conjectures, dogmas, and mystical beliefs have been put forward to explain the origin and nature of consciousness. None have gained general acceptance. But today, a compelling solution to the mystery of consciousness is at hand. The scientific disciplines, empowered by sophisticated technological tools, have identified the elements necessary to formulate a comprehensive theory of consciousness. The purpose of this book is to elucidate the new theory. In so doing, we will take an adventure on the *Anthropic Pathway*—the branch of evolution where consciousness originated and ascended.

Evolution on the anthropic pathway is characterized by the development of complex physical structures and the emergence of novel properties associated with those structures. In particular, life and consciousness are emergent properties associated with the most complex structures known.

The anthropic pathway has unfolded through seven phases of evolution. It arose from the symmetry of the big bang, fostered the development of process-structures into holistic living beings, and culminated with the formation of the conscious human brain:

1) Symmetry
2) Symmetry Breaking (physics)
3) Complexity (chemistry)
4) Holism (biology)
5) Symmetry Integration
6) Mental Consciousness
7) The Human Brain

Each phase of the anthropic pathway is described in the introduction of this book, outlining the evolution of the universe on a cosmic scale and highlighting how life and consciousness emerged and advanced on earth. Astronomical observations indicate that the universe has been expanding for 13.8 billion years and is pervaded by a homogeneous relic background radiation. These findings suggest the universe was once in a state of uniformity, thermal equilibrium, and absolute symmetry. But the timeline of evolution is asymmetric, proceeding from past to future only. The symmetry of time, as well as numerous other symmetries that specify the laws of nature, are reduced, or "broken," with respect to the underlying absolute symmetry of the universe.

The cause of symmetry breaking can be traced back—that is, deduced in terms of unidirectional time—to the incompatibility of global symmetry (pertaining to the whole) and local symmetry (pertaining to the parts). Global and local symmetries cannot be fully resolved into an absolute symmetry without introducing mathematical inconsistencies into the laws of physics, indicating a dichotomy, a primordial tension, inherent to the nature of the universe. This tension is relieved by the hierarchy of "gauge symmetry" laws descending from the big bang, which parametrize unidirectional time, the elementary particles, the gravitational, nuclear, and electromagnetic forces, entropic influences promoting order and self-organization, and, ultimately, complex adaptive systems evolving in realms of quantum Darwinism, natural selection, and neural Darwinism.

After the chemically rich planet earth had consolidated under the force of gravity, about 4.5 billion years ago, a tranquil environment hospitable to electrochemical interactions was established. Interaction pathways closed to form autonomous, self-reinforcing autocatalytic cycles that competed for limited chemical resources. The fittest cycles survived and evolved into *holistically complex* organic cells (holistically complex systems are unified, self-proficient entities possessing emergent properties distinct from those of its interdependent parts). These living, unicellular organisms were able to self-encode and utilize knowledge acquired from their prior experiences, thereby expressing the proficiency of *elementary consciousness*. For example, an amoeba that can seek out, recognize, and engulf paramecia, while continuously improving its survival skills on the basis of accrued knowledge, has elementary consciousness. Within colonies of unicellular organisms, the selective benefits of synergistic collaboration induced a second level of holistic closure, in the form of multicellular organisms, which expanded the capacities of elementary consciousness. However, these organisms lacked the means to be aware, mentally, of their experiences, for such an aptitude would require yet another level of complexification and holism—and the realization of *mental consciousness*.

The holistic multicellular platform provided an environment where cells adapted for the selective benefits of information production, transfer, and integration. In several types of primitive organisms, sensory-motor cells divided into discrete sensory and motor cells as well as interconnecting *principal neurons*. This laid the foundation for the simple nerve nets of animals such as the hydra (neurons are cells that exchange signaling molecules [excitatory or inhibitory neurotransmitters] across tiny clefts [synapses] separating the outgoing end [efferent axon] of a neuron from the incoming end [afferent dendrite] of another). Nerve nets were capable of transmitting signals from sensory cells to widely distributed motor cells via principal neurons, facilitating global reactions to local stimuli. In a subsequent stage of evolution, which transpired independently in a variety of species, a subset of the sensory cells further bifurcated to produce *interneurons*, the foundation for the more complex nervous systems of animals such as sea stars and worms.

Interneurons and short-range principal neurons form local microcircuits that communicate via long-range principal neurons. Microcircuits regulate the flow of signals through widespread logic and control systems, and coordinate the global synchronization of firing patterns. Specifically, microcircuits send out messages about local firing patterns to remote targets. These targets reciprocatively return messages, inducing the firing patterns of the local and distal structures to synchronize, or phase-lock, their oscillation frequencies, generating a spatial network of temporally organized, endogenously causal process-structures. In the most evolved neural networks, a spectrum of oscillation frequencies, called brainwaves, superpose constructively and destructively to promote, inhibit, and harmonize a multiplicity of interdependent functionalities, including emotional behaviors, cognitive processes, and attentional focus.

Interneurons are specialized to manage the integration of information, as well as to assemble complementary structures able to *collectively sense,* or *ultrasense,* that information. The collective sensation of integrated information is implemented exclusively within an emergent, novel realm of neuronal functionality, the *interneuronal domain,* where mental consciousness is made possible. (In vertebrae, the interneuronal domain is located entirely within the skeletal confines of the central nervous system, whereas the peripheral nervous system is absent of interneurons.) Moreover, the ultrasensory collective structure, which is assembled in concert with its complementary informational representation, is the entity that carries the highest form of experientiality, having the phenomenological essence that we realize in our waking and dreaming states of mental consciousness.

Interactions between collective structures and integrated information are fundamental to both consciously experienced and subliminal neural operations. For an informational representation to be consciously experienced, it must attain a certain threshold of stability. Experiments in cognitive psychology have correlated consciously reportable events with *sustained, reentrant* (recycling) states of informational representation; whereas brief or variable representations are always subliminal (although they may affect sensory-limbic, executive-motor, cognitive, memory, and other functionalities). Over extended durations, the reentrant cycling of information through a series of incrementally updated stable states has the effect of endowing mental consciousness with a coherent temporal flow.

The state of mental consciousness is produced through the process of *neuronal group selection* (also called *neural Darwinism*). The term "neuronal group" designates the minimal, indispensable neuronal structure able to support a particular, content specific, experiential state. To maximize synergistic benefits, the intricacy and characteristics of the experienced informational representation are regulated in conjunction with its complementary, experiencing, collective structure (together constituting the totality of the neuronal group), as well as nervous system performance and related functionalities. The process of neuronal group selection has evolved a repertoire of strategies, which includes: *resource allocation*—apportion limited neuronal resources to optimize overall (subliminal and experiential) efficiency and content; *uniqueness*—construct only one neuronal group at a time to ensure largescale consistency; *saliency*—experience only the most urgent, important, or relevant information; and *unitarity*—experience a coalesced, transcendent form of information enriched with quality and meaning. For example, when the sight of a person's face is perceived, its shape, color, and other "qualia," are always bound together and experienced as a whole. A unitary experience may encompass a wide variety of informational categories, qualities, semantic meanings, memories, and intentional impulses within a single representation. Hence, the sight of a certain face might invoke its recognition, related memories, feelings of emotion, an urge to act, and so forth, into a unitary experience.

Mental consciousness has three experiential components: the *experiencer* (the collective ultrasensory structure), the *experienced* (integrated information), and *phenomenal experience* (the essence of mental consciousness that supervenes upon the experiencer). The interface between the experiencer and the experienced consists of the group of dendritic arbor synapses that are active during an event of mental consciousness. (Dendritic arbors are densely populated, branching clusters of dendrites, ranging in complexity from the neuropil in the cerebral ganglia of a primitive sea star, to those that foliate the pyramidal neuron/interneuron framework of the human neocortex). Additionally, dendritic arbors are the repositories for memories encoded in the configuration of synaptic spines which stud the surface of each dendrite.

The collective structure of the experiencer consists of a patterned array of post-synaptic dendrites, and the informational representation it experiences is rendered by an inversely patterned array of pre-synaptic axons. Signaling between the experiencer and the experienced is conducted both through and around the core synaptic interface. Feedforward signals transmitted from the pre-synaptic side of the interface bypass it in order to shape the collective structure directly, priming it to interact with only the most salient and contextually relevant informational signals passing through the interface. In turn, feedback signals transmitted from the post-synaptic side of the interface cycle back and reenter the pre-synaptic input stream, prioritizing, filtering, and qualifying the very (sustained, recycling) information being experienced by the collective structure. Thus, the visual image of a cat (pre-synaptic signaling pattern originating in the retina) might induce (based on the form of the post-synaptic collective structure) an experience of fondness in a human being but one of fright in a mouse.

In a highly evolved neuronal group, the signaling pathways traversing the core synaptic interface are massively parallel, bidirectional, and profusely interconnected, establishing an architecture based on coupled and nested cycles. At this level of complexity, the physical distinction between the experiencer and the experienced is inscrutable. Nevertheless, while the core interface may be subsumed in the complexity of the neuronal group, the reciprocative duality between the experiencer and the experienced retains its archetypal, causative influence on the third component of mental consciousness—phenomenal experience—which has its own, subjective, ontological nature.

Mental consciousness reflects two ontologies: The first-person, subjective, directly known ontology refers to experienced information in general; it stems from the endogenously projected perspective of the collective structure toward integrated information. The third-person, objective, indirectly surmised ontology refers to the specific informational content derived from empirical observations and inferences about the systems, such as the brain, which have been correlated with experiential events.

Longstanding attempts to explain experientiality in terms of the third-person ontology alone have failed to solve the "mind-body problem," i.e., how to bridge, via a continuous causal chain, the "explanatory gap" that separates the physical brain from the phenomenal mind. This gap is illusory, however; dispelled by recognition that both the first- and third-person ontologies refer to the same physical state. The one physical state *is* the bridge that connects two perspectives. The bridge does not connect the physical brain to a distinct experiential mind; rather, it connects the third-person, objective ontology conjectured about a physical state, to the first-person subjective ontology that is based on that state's experiential essence.

* * *

Part 1 of this book describes a theory about the *Evolution and Nature of Consciousness*. In Chapter 1, the foundation for the theory is developed in terms of basic symmetry principles. In Chapter 2, the concept of symmetry spaces is extended to encompass factors of complexity and holism. Chapter 3 examines selection and self-organization, and Chapter 4 identifies the origin of life on earth and how its evolution parallels the infoldment of the complex process-structures by which organisms produce integrated information. Chapter 5 reviews the specialization of the nervous system and its informational basis. Chapters 6 and 7 describe the extremely complex neurophysiology of the human brain as being an instrument capable of integrating information and experiencing it endogenously from a first-person orientation. Chapter 7 turns inward, discussing self-awareness, meditation, and the experiential essence of mental consciousness which supervenes upon ultrasensation. In Chapter 8, consciousness is unfolded, providing an explanation for its origin and evolution. Chapter 9 considers other perspectives on the nature of consciousness, including psychology, Eastern and Western philosophies, and theology.

Part 2, *Computers and Artificial Consciousness,* is about artificial intelligence machines becoming conscious. Chapter 10 provides an overview of electronic computer technology, emphasizing the application of the object orientation paradigm that is intrinsic to living systems. Chapter 11-1 describes how deductive logic limits knowledge of reality and how it must be transcended by cognitive logic, which governs brain operations, in order for a machine to be conscious. In Chapter 11-2, an original graphical proof of Gödel's Incompleteness Theorem is presented to demonstrate that deductive logic is an insufficient foundation for self-referential systems, including lifeforms and the brain. Chapter 12 reviews the history and technology of artificial intelligence machines, and identifies the essential requirements for a neuroevolutionary machine to attain holistic closure and the emergence of consciousness.

Part 3, *Physics and Metaphysics,* is presented for readers interested in the more technical principles of theoretical physics and the deeper ontology underlying the cycles of the cosmos and evolution along the anthropic pathway. Chapter 13 investigates gauge theory, describing how global and local symmetries are mediated in the standard model and relativity. Chapter 14 describes symmetry-group theory, and the many types and characteristics of the symmetries enfolded in nature. Chapter 15 studies string theory and M-Theory, cyclic braneworld cosmology, and the origin and nature of entropy. In Chapter 16, the ubiquity of cycles in nature is demonstrated, ranging from loop quantum gravity, to the inter-cycling of environmental-biological-neural pathways, to the life-cycle of the universe. Chapter 17 discusses quantum mechanics in detail, comparing its orthodox interpretation to David Bohm's ontological interpretation, which "emphasizes the wholeness that is signified by quantum laws." In Chapter 18, the anthropic pathway is retraced in terms of astrophysics; quantum Darwinism, natural selection, and universal Darwinism are reviewed; and the "fine tuning problem" is investigated—i.e., what was the cause or reason for the universe to have possessed the improbable coincidence of property values necessary for complexity and conscious life to emerge and evolve? The multiverse solution to the fine tuning problem is negatively critiqued; instead, a specific evolutionary pathway that has developed within the event horizon of our observable universe is delineated. This pathway has traversed an extremum between order and disorder, proceeding inexorably toward an anthropic attractor of complexity, and the conscious human brain.

* * *

Introduction

THE ANTHROPIC PATHWAY

1) *SYMMETRY*

The ancient Greeks used the term *summetria* (symmetry) to designate geometrical objects having the same shape. Thus, every circle, regardless of its diameter, is said to *represent* a particular symmetry (roundness). In modern times, the concept of symmetry has been expanded to include the laws of physics. For example, Newton's law of force, F = ma (Force = mass x acceleration), designates a symmetry because, regardless of an object's mass or acceleration, the mathematical *form* of the law (F = ma), like the geometrical *shape* of a circle, never changes.

Collectively, all of the states a system can possibly occupy comprise its *symmetry space*. A simple example of a symmetry space is provided by a lightbulb which can be either on or off; hence, this symmetry space encompasses two possible representations. Other symmetry spaces found in nature may be much larger and more intricate. The entire macrocosmic universe is a symmetry space whose representations (possible configurations of the universe) change in accordance with the laws of cosmology; and, looking ahead, the human brain comprises a symmetry space wherein informational electromagnetic patterns are represented in accordance with the laws of neuroscience.

2) *SYMMETRY BREAKING*

Evidence from particle physics and cosmogony suggests that numerous symmetries currently observed in the universe are "broken" from a state of absolute supersymmetry, in which time itself is resolved. The reason for the supersymmetry of the universe to be broken is that the global and local effects of quantum fluctuations, the fabric of the cosmos, cannot be resolved with mathematical consistency into a supersymmetry. Thus, the supersymmetry of the universe is broken into symmetry groups; laws of physics that mandate a unidirectional timeline (which emanated from an asymptotically reckoned event called the "big bang"), macrocosmic expansion and cooling, the elementary particles and their interactions, and an entropic force conducive to complexity and self-organization.

3) *COMPLEXITY*

In the early, cooling universe, the elementary particles produced through symmetry breaking were able to bond into atoms and molecules. These composite structures exhibited emergent properties and capabilities not possessed by any of the constituent parts. For example, certain molecular formations, called catalysts, could activate particular chemical reactions in favor of others, resulting in networks of reaction pathways and cyclic interdependencies, or *complexity*. Complexity thus dictated the utilization of environmental resources in a manner that *selected* for the fittest of systems.

4) *HOLISM*

Four billion years ago, inside tiny chambers that riddled the walls of deep-sea vent structures, the complexity of catalytic systems advanced dramatically. (The deep-sea vent structure is chosen to describe the kind of environment in which life emerged. Other settings have been proposed to have provided the necessary conditions where abiogenesis actually occurred: Darwin's little pond, lipid protocells, volcanic hot springs, etc., all employ the same principles of selection, complexification, self-organization, and autocatalytic closure delineated below, heuristically, in terms of deep-sea vents.) The protective chambers within the deep-sea vents were tranquil inside, though flooded with chemicals entering from the earth's churning mantle below and from the surrounding ocean waters. An extensive variety of large molecules, new catalysts, and self-modulating feedback and feedforward cycles evolved through continuous selection for complexity. This enhanced the overall strength and versatility of the intra-dependent chambered system such that, eventually, extremely complex chemical networks, termed autocatalytic sets, achieved *holistic closure,* marking the advent of life.

Holistically closed living systems possess three vital characteristics: *self-proficiencies, biosemiosis, and experientiality.* The *self-proficiencies* include self-referentiality, self-organization, self-renewal, and self-replication (whereas inanimate objects such as rocks or computers are not holistic because they lack the complexity to implement self-proficiencies). Self-proficiencies are based in holism and are governed by causal forces derived from that holism. These holistic forces represent laws of nature which are manifested in the symmetry spaces of living systems, or microcosms, in emergent dimensionalities that complement those of basic chemistry. After holistic closure was attained in vent chambers (or similar setting hospitable to the formation of stable polymer strands billions of molecules long), specialized process-structures called organelles—such as ribosomes to synthesize bodily materials, mitochondria to produce energy, and membranes to regulate the environmental interface, along with DNA-based control centers to catalyze the production of organelles—all consolidated and worked cooperatively within the holistic symmetry space of the "organic cell."

Living systems are *biosemiotic.* They require a flow of materials and informational signals, called "signs" in biosemiotics, to pass through their process-structure, referred to as the "interpretant." Biosemiosis signifies a critical enhancement to the dyadic cause and effect relationship that configures the interactions of classical, non-complex, physical systems. In contrast, the causal architecture of a living, complex, open-system microcosm is triadic, consisting of the conjunction of afferent (incoming) and efferent (outgoing) sign streams, and a mediating, causally affective interpretant process-structure. (An open system has throughput and variable contents, as opposed to an isolated-system in which contents are conserved. A living biosemiotic system, while holistically closed, is an open system.) Thus, a cell, a tree, a sea star, and a human being are all holistically-closed, open-system, biosemiotic interpretants processing afferent and efferent sign streams within their holistic symmetry space. For example, the unicellular green alga *Chlamydomonas* has eye-apparatus capable of sensing light (afferent signs), processing these signs (interpretant), and then directing its motor flagella to move by broadcasting intracellular electric signals (efferent signs).

As defined by Webster, *experience* is "the direct participation in or the observation of events as a basis of knowledge." A unicellular organism like an amoeba, which can sense and recognize a paramecium, and decide to engulf and assimilate it based upon the content of its self-embodied knowledge base, satisfies this definition of experience. Similarly, a tree, sensing light, moisture, gravity, or pressure in its environment, can process and integrate that information, evaluate it with respect to its stored knowledge base, and determine the optimal growth trajectories through which to project its roots, altogether constituting an experience. New experiences may be encoded into the process-structure of the tree to update its knowledge base, which may later be called upon to reiterate or avoid that experience.

An experience realized by any holistic biosemiotic living system may result either from signs that originate in the external environment—such as the experience of hearing a bell—or from signs produced within the system's internal milieu of interdependent pathways—such as the experience of a thought. Whether stemming from external or internal sources, experiences update the process-structure and knowledge base of an organism, and influence the nature and consequences of its subsequent experiences.

5) *SYMMETRY INTEGRATION*

In primitive deep-sea vent chambers, information processing concerned only the regulation and balance of chemical concentrations that were spread out in space and unsynchronized in time. Although information was integrated through the holistic interconnectedness of the chemical pathways—constituting an integrated symmetry space—there were no mechanisms to construct focused informational patterns. It would take the selective powers of *infoldment* (the concentration of process-structures in space and time for improved efficiency and cohesive specialization) and *synergy* (the working together of process-structures for the benefit of the whole) to partition and integrate informational patterns into configurations propitious for selective advantage. As life systems evolved, pathways and cycles consolidated into functional units (e.g., the organelles and DNA control centers mentioned above) able to recognize and produce higher-level patterns of integrated information.

Evolving microcosmic symmetry spaces have ascended a hierarchy of informational platforms as new levels of biosemiotic holism, each supporting its own self-proficiencies, have been realized through repeated cycles of differentiation (specialization of individual parts) and integration (synergistic interconnectedness of parts). First, there was the synergistic specialization by which colonies of molecules became the organelles of holistic unicellular organisms. Next was the synergistic specialization by which colonies of cells became the organs of holistic multicellular organisms. Then, and most recently, the synergistic specialization by which colonies of neuron cells, dedicated to informational relay (principal neurons) and integration (interneurons), became the building blocks of holistic mental nervous systems, the platform upon which the most advanced attributes of consciousness have been achieved.

In general, neurons have the following characteristics: Neurons interchange signals via electrical oscillations in their membrane potentials, or by emitting and reacting to electrochemical ions. The electrochemical signaling currency of neurons has taken the form of small *neurotransmitter* molecules. Neurotransmitters evolved from large prohormone-derived signaling peptides used in non-neuronal signaling systems. A neuron accumulates voltage through its afferent connections, or *dendrites,* located on its input end, and releases neurotransmitters from efferent connections on its output end, or *axon,* upon the accumulation of a specific *action potential* voltage on the cell body. Neurotransmitters may induce or inhibit other neurons to reach their particular action potential firing thresholds. Action potentials were originally involved in non-neuronal mechanisms such as flagellar beating in motile single-celled choanoflagellates, which later colonized to form multicellular organisms that inherited and exapted the action potential underpinnings. Neurotransmitters are conveyed across tiny clefts, called *synapses,* which are located between the efferent axon of one neuron and the afferent dendrite of another. Synapses were co-opted from ancestral peptide-based proto-synaptic devices utilized in cell-to-cell communication among unicellular organisms, including colonial choanoflagellates.

The first nervous systems were in the form of diffuse nerve nets located just beneath the epidermal layer of certain marine organisms. Nerve nets facilitated prolonged, coordinated, and integrated behaviors. Nerve nets appeared independently in the phylogenic branches of *Ctenophora* (comb jellies) and *Cnidaria* (hydra, jellyfish, sea anemones, and corals). Within evolving nerve nets, infoldment led to the clustering of groups of neurons into ganglia, nuclei, and centralized proto-brains. Later, *Bilateria* split from the *Cnidaria* line, and it is in this line that the most extreme degrees of neural complexification, anatomical and functional specialization, and informational integration have been attained, notably in the neocortex of *Homo sapiens.*

The nervous system is often considered, or defined, to be the only platform upon which consciousness exists. However, the systematic evolutionary process of symmetry space integration commenced with the first holistic lifeforms, making the inference that consciousness is a general potentiality of biosemiotic holism both sensible and constructive. Returning to the example of a tree; the tips of its roots, the "radicles," can sense light, gravity, moisture, and pressure, as well as various toxins, microbes, and chemical signals coming from neighboring plants. Based on detected signs, growing roots about to encounter an impenetrable object or a toxic substance can change course before making direct contact. The tree can determine whether nearby roots are its own, its kin, or another species of tree. This demonstrates that a tree has an internal awareness of its surrounding conditions, derived from the process-structure and knowledge base of previous experiences. We now define this type of awareness as *elementary consciousness*, and the domain of its internal implementation as its *experiential symmetry space*. However, the tree is not aware, mentally, of the obstacle or toxin in its path, for such an aptitude requires the more advanced faculty of *mental consciousness*.

6) *MENTAL CONSCIOUSNSS*

Mental consciousness arose from the specialization of sensory-motor cells in several types of primitive nervous systems. In certain ancient marine organisms, mechanosensory-contractile cells located on the epithelial surface—the "founder cells" of the nervous system—developed a synergistic protective mechanism that prevented the influx of the lethal calcium ions saturating the seawaters of 1.5 billion years ago. These mechanosensory-contractile cells adapted into three collaborating cell types to harness the efficiency benefits attainable through division of labor. First, the mechanosensory-contractile cell divided into distinct epithelial mechanosensory and contractile cells that communicated through synaptic conduction, laying the foundation for the simple nerve nets of primitive animals such as the hydra. These nerve nets were capable of transmitting signals from mechanosensory receptor cells to widely distributed contractile effector cells, which facilitated global contractions and locomotive movements. A subset of the mechanosensory cell then divided into a mechanosensory principal neuron, and an interneuron specialized for the integration of information and the assembly of collective structures able to process that information. The emergence of interneurons instituted a new realm of functionality within the microcosm, the *interneuronal domain*.

Following the same evolutionary paradigm undergone by the mechanosensory-contractile cell, subsets of both the photoreceptor and pigmented shading cell types in the progenitor of the simple-eyed annelid *Platynereis dumerilii* specialized for the benefits of intercellular division of labor and collaboration. These cells relinquished their photosensitivity and shading capabilities, transforming into interneurons specialized for integrating modes of visual information. As evolution progressed, increasingly complex and integrative interneuronal domains arose. Within the nerve nets of more evolved species of hydra, interneurons specialized to integrate information from ciliary-sensory tactile and photoreceptive cells for the purpose of coordinating the actions of its epithelio-muscular cells. Initially, the evolutionary benefit of inducing the integration of sensorially- and endogenously-produced information within an organism was the improvement of its survival competencies with respect to the external environment, by providing it with an expanded knowledge base with which to better select and implement its motor/behavioral actions. Subsequent to the formation of these primitive, diffusely organized interneuronal domains, spatial centralization further enhanced process efficiencies and bodily coordination, as seen in the centralized cerebral ganglia of the radiata sea star *Asteroidea,* in the mushroom body brains of insects such as the bilateria fruit fly *Drosophila,* and, ultimately, in the skeletal encasements of the central nervous systems of the vertebrae, including *Homo sapiens,* where the pinnacle of mental consciousness has been realized.

The evolution of the interneuronal domain has been characterized by the emergence of a variety of interdependent properties and features, three of which are essential elements of mental consciousness: collective sensation of integrated information, complexity and holistic closure, and subjectivity and the phenomenal symmetry space. We now describe each of these requisite features of mental consciousness in turn:

Collective sensation of integrated information. Upon genetic underpinnings exapted from unicellular sensation, the multicellular colonization and networking of cells within the interneuronal domain led to the holistic closure of an interpretant process-structure able to *collectively sense* integrated information. A given occurrence of collective sensation is complementary to a specific representation of integrated information, which is constituted within the interpretant process-structure itself (the physical relationship between a collective sensory structure and the informational representation which it senses will be elucidated below). This emergent collective form of sensation will be called *ultrasensation* when referring to the nascent potentiality that arose upon the bifurcation into and the re-association of sensory neurons and interneurons. But when referring to collective sensation in its more evolved form—characterized by sustained, reentrant cycling—the term *mental consciousness* will be used.

There is a fundamental difference between basic sensation as possessed by a unicellular organism or a multicellular organism having a simple nerve net, and the collective ultrasensation-mental consciousness of an organism possessing an interneuronal domain. Basic sensation entails a dyadic cause and effect relationship whereby signals from sensory cells are transmitted directly to effector cells. In contrast, ultrasensation-mental consciousness, involves a triadic biosemiotic causal architecture, incorporating an interpretant process-structure that mediates between basic sensory and effector systems.

Mediation of sensory-effector signaling by a process-structure that can implement collective sensation of integrated information improves a living system's ability to flourish in its external environment. Collective sensation of integrated multimodal information offers an organism increased survival advantages by providing it with a larger, more comprehensive grasp of its surroundings, and consequently affords it competency to more effectively respond to environmental challenges and opportunities. Complementing selective fitness achieved through the collective sensation of integrated multimodal sensory information, the incorporation of information about an organism's internal conditions (acquired by projecting neuronal cycles from the interneuronal domain into tissues and functional systems throughout the body) further extends the organism's scope of collective sensation and overall fitness. Information about an organism's physiological state enables it to fine-tune its homeostasis (the maintenance of internal conditions necessary for survival). In addition to homeostasis, ultrasensation of internal conditions initially concerned instinctual information related to fulfilling the most basic needs of an organism, such as nutritional sufficiencies or responsiveness to genetically expressed replication impulses. Moreover, the interpretant process-structure may operate independently of the sensory, motor, and homeostatic systems lying outside the interneuronal domain by processing and cycling sign streams flowing exclusively within the domain, as in the case of the cycles encoding and recalling multimodal memories, and, in advanced domains, for the logic cycles facilitating cognitive thinking. Furthermore, selective partitioning of the interpretant network into a unitary, reentrant, and sustained ultrasensory complex establishes the endogenous environment underlying the phenomenal symmetry space of subjective experientiality.

Complexity and holistic closure. The interneuronal domain is not merely a network that relays patterns of integrated information for collective sensation and mechanical response. The interneuronal domain has evolved into a feedback-feedforward network of such complexity that it has attained holistic closure. Previously described in the sections entitled *Complexity* and *Holism,* the interneuronal domain has evolved by means of the same principles of selection for complexity that had engendered the unification of holistic unicellular organisms from catalytic sets, and holistic multicellular organisms from colonies of unicellular organisms. Although the complexity of any holistically closed system is intractable, a coalition of interdependent process-structures and functionalities in the interneuronal domain that correlates with events of mental consciousness has been identified:

Interneurons form local microcircuits that communicate, via principal neurons, across widely distributed areas of the interneuronal domain. This regulates the flow of signals that support large-scale logic and control systems, and the synchronization of global firing actions. Notably, across a special type of interneuron-rich neuronal tissue called cortex, this spatiotemporal coordination binds distributed informational patterns into integrated unitary representations. Included within the informational representations produced in the human cerebral cortex are recallable memories previously recorded into the very cortical structure that represents them. At the comprehensive level, memories engage signaling pathways interconnecting cortical areas dedicated to different informational modalities, e.g., tactile, visual, or abstract. At the granular level, within a dedicated cortical area, interneurons coordinate interactions for a large variety of specialized types of neurons. This includes a type of principal neuron, called a pyramidal neuron, where memory data is encoded in the arrangement of synaptic spines located on the surface of its dendrites—pyramidal neurons have a cell body shaped like a pyramid, a projecting axon, and densely foliated apical (afferent) and basal (efferent) dendritic trees. Each dendrite in a tree is profusely studded with both excitatory and inhibitory synaptic spines, resulting in tens of thousands of neuroplastic connections per neuron. Moreover, dendritic trees intermesh to form memory-shaped dendritic arbors that function as cooperative process-structures.

A vital potentiality of the interneuronal domain that must be realized in order for mental consciousness to be effectually manifested is *reentry*, the recursive bidirectional exchange of signals along reciprocal axonal fibers connecting two or more cortical areas. Experiments in cognitive neuroscience, employing advanced techniques of magneto- and electro-encephalographic imaging, applied to the human cerebral cortex, have shown that there is significant variability in neuronal firing patterns during subliminal perception tasks. In contrast, there is a pronounced decrease in variability for consciously reportable tasks, demonstrating that *sustained, stable, reentrant cycling* within the cortex is a signature of mental consciousness. Indeed, the portion of the interpretant network that is linked together into a sustained, stable, complex of reentrant cycles demarcates the symmetry space wherein instances of mental consciousness are realized.

The interneuronal domain has been the site for the evolution of many holistically beneficial neural functionalities. Long-range spatial interconnectivity and widespread temporal synchrony, a self-embodied knowledge base, integrated and collective states, and the faculty of mental consciousness, are products of the complexification that has taken place in the interneuronal domain. All of these interdependent functionalities are synergistically unified within an overarching hierarchical network by which the interneuronal domain is organized. The hierarchy is comprised of localized process-structures, called *nodes*, which are interconnected through bidirectional pathways traversing and spanning the levels of the hierarchy. The nodal network will be further explicated in the next section where the neural architecture underlying human mental consciousness is explored. For now we may envision the levels of the nodal network as a set of concentric rings, separated by one synaptic step, corresponding to the levels of the hierarchy, with each ring beaded by function-specific processing nodes—for example, in the human brain, the primary auditory cortex (auditory function) and the limbic cortex (emotional function) are nodes. Arcs between nodes in a ring represent bidirectional connections between them, and the set of all nodes and arcs in a ring represents a collective structure capable of ultrasensation.

Nodes located in different rings are also connected by bidirectional pathways that transmit informational signal patterns formatted apposite to the level of integration compatible with the collective structures with which they interact. Nodes receive afferent signs from nodes in other levels, called "bottom-up" signs in the context of signs ascending through the network's integrative hierarchy; and transmit efferent signs based upon this integrated information, called "top-down" signs in the context of signs both descending through the action/behavior-inducing hierarchy and through the contra-directional return-cycle-arc of the *(sic)* afferent hierarchy. Top-down regulation of activities through the nodal network is not simply a reflexive capacity to process and transmit signals. The interneuronal domain calls upon its hierarchical nodal process-structure and self-embodied knowledge base to manage the informational signals which may enter into, ascend and descend through, issue from, and cycle within the domain, thus performing as a holistic executor of endogenous functionalities. Endogenous control, of functionalities and information processing in the interneuronal domain, endows it with a self-centered orientation, the basis for the first-person, subjective perspective of mental consciousness.

Subjectivity and the phenomenal symmetry space. Mental consciousness entails not only the physical systems described above, but also the proficiency of *mental experientiality*, i.e., the subjective awareness of an instance of integrated information produced by those physical systems. Before examining the experiential aspect of mental consciousness in more detail, operational terms are defined:
Subjective: existing in the mind (the domain of mental consciousness), from a first-person perspective.

Phenomena: things as they appear to and are constructed by the mind; phenomena are the appearances which constitute the subjective content of mental experiences. Phenomena have a qualitative character, or *qualia,* e.g., the *feeling* of an emotion, the *taste* of a strawberry.
Phenomenal symmetry space: the set of all potential phenomena that may appear in the mind.
Objective: existing independently of the mind of a subjective observer; i.e., the presumed or inferred reality of third-person objects and relationships in the external environment and within the physical body of a microcosm.
Noumena: things which in themselves are inaccessible to the mind and to which phenomena are referred. Noumena are presumed or inferred things onto which we impose our concepts, knowledge, and beliefs; e.g., physical objects in the external world.

In an event of ultrasensation, an *informational signal pattern* interacts with a collective structure that contextualizes the pattern so as to produce an *informational representation.* This representation is ultrasensed by the very same collective structure that conjoins with the signal pattern in the interaction. When an event of ultrasensation is sustained, reentrant, and unitary, the collective faculty of mental consciousness is realized. Mental experientiality is associated with the endogenous properties of mental consciousness, supervening upon its first-person orientation toward informational representations. Thus, the term mental consciousness has a more comprehensive meaning than mental experientiality; it refers to relevant representations of integrated information, collective process-structures, first- and third-person perspectives, as well as mental experientiality. Mental consciousness entails an involuted process-structure which will be unraveled as we proceed.

The conformation of neurons that delineates the interface between an informational signal pattern and the collective structure with which it interacts determines the informational representation, i.e., the subjective content, of a mental experience. The terms "informational signal pattern" and "informational representation" distinguish two kinds of information involved in an event of mental consciousness. Informational signal patterns (Shannon information; see Chapter 7) do not directly encode qualitative meaning relative to an experiential structure, while an informational representation (semantic information), formed by the conjunction of a signal pattern and an evolved (sustained, reentrant) collective structure, confer a non-trivial phenomenal, first-person quality of meaning relative to the structure. Hereafter, unless specified explicitly or contextually, ultrasensory structures which are enhanced by sustained, stable, and reentrant cycling—that is, vehicles of mental consciousness—are presumed.

There are three experiential aspects of an event of mental consciousness: experience, experiencer, and experienced. *Experience* is the qualitative, subjective appearance of an informational representation relative to a collective neural structure. The *experiencer* is the collective structure and the *experienced* is the integrated information. Experienced information is the product of the conjunctive interaction between the collective structure and signal patterns coming from various cortical and sub-cortical sources. The collective structure itself contributes a semantic dimension to the information it experiences, which is derived from its self-embodied knowledge base built through previous interactions. This semantic component of information is encoded in the conformation of the collective structure's dendritic arbor; i.e., the experienced informational conjunction unifies Shannon informational signals cycling in from external and reentrant process-structures with semantic signal qualifiers embodied within the collective structure's interfacing configuration. Thus, the conjunctive interaction engenders information that caries meaning and salience relative to the experiencer; for example, the same visual image of a cat (the signal pattern from the retina of the eye) might evoke an experience of fondness in a human being but one of fright in a mouse.

The boundary between the experiencer and the experienced is the set of dendritic arbor synapses active during an event of mental consciousness. Dendritic arbors range in complexity from the neuropil of primitive cerebral ganglia in a sea star to those that foliate the juxtaposed assemblies of pyramidal neurons and interneurons comprising the microcolumns that cross through the six layers of the human neocortex. Dendritic arbor synapses separate the pre-synaptic axons constituting the signaling structure that contributes to experienced information, from the post-synaptic dendrites constituting the input structure of the experiencer. Both the pre-synaptic (informational signaling) structure and the post-synaptic (experienced-experiencer) structure are grouped structures, each of which acts as a unified entity. Pre-synaptic axons (whose threshold discharges are dependent on voltage boosts delivered by post-synaptic reentry cycles) trigger apposite post-synaptic dendrites (which encode relevant memory engrams in their distributed but spatiotemporally connected structure) to fire collectively and recursively; i.e., to fire as a unit expressing an endogenous, self-centered orientation toward the semantic informational representation that it fashions; to function as a unit expressing sustained ultrasensation—that is, mental experience—of that information.

The properties and interactions of the neural systems empirically correlated with events of mental consciousness on the basis of objective, third-person observational techniques are the very same properties and interactions that are the basis for the first-person, subjective aspect of mental experientiality. An instance of integrated information and the collective structure to which it refers is a unified physical construct, but may be viewed from two epistemological perspectives: a subjective, first-person perspective based on direct phenomenal experience, and an objective, third-person perspective based on inferences (posited and comprehended in the first-person) about the (putative) noumenal world. Of course, perspectives of third-person noumena are ultimately known and experienced as first-person phenomena. For instance, the image of a cat may be directly experienced subjectively as a first-hand mental experience, but it may also be described objectively in terms of the neuronal firing patterns produced in the brain which correlate with that experience—both perspectives refer to the same cat, both are valid, but ontologically different. Specifically, certain states that are active in the interpretant process-structure during an event of mental consciousness have both an objective, third-person, indirectly surmised ontology (the empirically detected electrical firing patterns that arise in the mouse's brain when it sees a cat), as well as a subjective, first-person, directly known ontology (the conscious awareness of the cat by the mouse).

A historically notorious philosophical problem that persists to the present day has been resolved by the theory of mental consciousness presented here. This problem is usually referred to as the mind-body problem or the hard problem of consciousness, which seeks to identify the bridge that crosses the so-called "Cartesian divide" or "explanatory gap." Where, what, or whence is the transition or interface between the physical systems of the body and the *(sic)* nonphysical property of phenomenal experientiality?

In Western science and philosophy, attempts to bridge the explanatory gap have generally been framed in terms of a third-person ontology based upon empirical observations. Likewise, we have traced the evolution of process-structures, integrated information, and vehicles of consciousness on the anthropic pathway, and identified the physical interface between informational and ultrasensory structures with respect to a third-person causal ontology. But phenomena and subjectivity are characterized by a first-person ontology as well, which stems from the self-centered vantage point of an endogenous process-structure in the interneuronal domain. The explanatory gap has not, and cannot, be bridged in terms of a third-person ontology alone. The first- and third-person ontologies refer to the same physical state. This one physical state, conjoining informational and ultrasensensory structures, *is* the bridge. However, the bridge does not connect physical informational representations in the brain to a distinct entity called the mind, but rather, it connects a third-person, objective ontology with a first-person, subjective ontology upon which the property of experientiality supervenes.

The subjective ontology of mental consciousness is multifarious, yet unitary, elaborate, and prolific. The diversity of phenomena experienced through human mental consciousness includes numerous informational types and qualities. The experiential repertoire encompasses integrated combinations of basic sensational qualia (the redness of red, the feeling of pain), emotions (anger, love), meaningful constructs of the intellect (the notion of justice), cognized logical flows through time (a mathematical computation, the thread of a conversation), imaginations, and, importantly, discernments of intentional projections from the first-person mental self toward the third-person noumenal world, both in the form of executive-motor actions directed outward to the external world, and top-down projections directed within the world of mind and body. These and all other varieties of subjective, first-person phenomena are often denoted comprehensively by the term "qualia," a convention we shall adopt.

Cortical nodal regions correlated with purposive intention, in its various functional forms, such as attention and planning, have the same neurophysiological and causal structure as regions correlated with the bottom-up functionalities of mental consciousness, such as visual perception and language comprehension. All regions, whether associated with sensory, executive-motor, or abstract functionalities, possess dendritic arbors shaped through experience, as well as a harmonized pre-synaptic informational structure and a collective post-synaptic structure capable of ultrasensation. It must be emphasized that intentionality is paramount to the nature and power of human mental consciousness, for it enables us to secure sovereignty over our minds and bodies.

* * *

The broad conception of consciousness outlined above, encompassing elementary and mental consciousness, is both practical and constructive. It naturally positions the origin of elementary consciousness at holistic microcosmic closure, delineates its step-by-step evolution through phases of complexification and symmetry space integration into neural vehicles of mental consciousness, and it identifies the subjective phenomenal symmetry space of mental consciousness as an emergent self-property arising upon the faculty of collective sensation. Given that all holistic microcosms possess some form of consciousness, we might ask: What is consciousness like for a unicellular amoeba, a multicellular tree, or a sea star possessing only a rudimentary nervous system?

Numerous phylogenic strategies have found success on the anthropic pathway. The great range in informational focus and the richness of conscious experience, varying from vague asynchronous impressions in dispersed chemical sets, to vivid multifaceted mental awarenesses in the human brain, all point to the diversity of life, and to the scope and essence of consciousness. This intriguing topic will be explored in the chapters following the Introduction, where it is shown that the objects of consciousness that characterize all holistic symmetry spaces differ essentially in terms of their complexity in space and time, the types and levels of integrated information they represent, and the modes of experientiality—elementary and/or mental—associated with them. Thus, while all forms of consciousness provide vital selective benefit to an organism, the focus and richness of the objects of consciousness they experience should not generally be ascribed with the remarkable features that are unique to the human brain, the most complex structure we have yet encountered in the universe.

7) *THE HUMAN BRAIN*

In the human nervous system, the interneuronal domain has infolded into an open-system symmetry space termed the *central nervous system.* The central nervous system includes the spinal cord and the brain, housed within the vertebral column and the skull, respectively. The central nervous system consists of principal neurons as well as all of the body's interneurons, whereas the exterior peripheral nervous system consists of principal neurons only. The central nervous system demarcates an endogenous environment twice removed from the external noumenal world: first, by the horizon boundary separating the external macrocosmic environment from the internal open-system symmetry space of the holistic living microcosm; and second, by the horizon boundary separating *that* microcosmic symmetry space from the included open-system symmetry space of the central nervous system, wherein transformations and functionalities of the emergent integrative interneuronal domain are expressed.

The new types and levels of interdependencies that have evolved on the platform of interneuronal connectivity underlie the numerous and advanced functionalities which differentiate the capacities of the human brain from those of all other species. To optimize these functionalities, the evolved brain has partitioned into two sectors: the cerebral cortex and a large group of complementary subcortical systems—during ontogenesis (prenatal development) a neural tube develops along the axis of the central nervous system and partitions into five regions, called encephala. The anterior region (at the top of the axis) is the telencephalon which subdivides into the cerebral cortex and the basal ganglia (involved in motor, reward, and executive functions), while the other four regions develop into subcortical structures including the thalamus (a group of ~40 nuclei, some of which perform as gateways between the external world and the functional areas of the cortex, while other nuclei are involved in emotions, thinking, and conscious alertness), the amygdala (emotions and attention), and the brainstem (bidirectional conveyance of sensory and motor signals between the brain and body). See Chapters 6 and 7 for illustrations and further descriptions of brain anatomy.

The developed human central nervous system may be visualized schematically as a series of structures ascending along its cephalization axis—cephalization is the evolutionary trend for functional structures having the highest degree of informational integration and executive jurisdiction to be positioned toward the top of the central nervous system axis, whereas local functions are implemented lower on the axis. Rising from the bottom of the axis are the spinal cord, the brain stem, the thalamus, and, at the top, the cerebral cortex. Signaling between the thalamus and the cortex employs different schemes for sensory input and motor output. Sensory signals radiate from nuclei in the thalamus directly to functionally corresponding areas in the cortex. In contrast, motor signals from cortical areas converge to the basal ganglia which processes them and forwards modulatory signals to the thalamus, which, in turn, sends signals back to the cortical area from which they originated, closing the basal ganglia-thalamocortical loop. This fundamental pathway structure of the central nervous system has been conserved throughout vertebrate evolution from the time of the primitive lamprey, 560 million years ago.

The cerebral cortex is generally homogeneous in composition, consisting of modular columns about 0.5 mm in diameter which pass through the layers of the cortex totaling about 2.5 mm in thickness. The basic arrangement of functional areas across the cortex is established through genetic expression during ontogenesis, but the modular columnar composition facilitates significant plasticity in later stages of life, so that functional areas may expand, contract, move, coopt, and share resources. But neural resources available in the cortex are limited, and therefore the priority, scope, speed, and amount of information that can be practicably integrated for a particular function are constrained. Accordingly, to ensure optimal holistic synergy, resources apportioned to the function of mental consciousness are themselves strictly regulated and moderated, with competing routine and contextually less salient tasks off-loaded to parallel subliminal, peripheral, and autonomic systems.

In the cerebral cortex, signals flow bidirectionally through the nodal network, mentioned earlier. A node consists of a grouping of neurons arranged in a column that passes through the cortical layers. A column typically spans a few square centimeters in area. Functionally, the nodal network is organized in a hierarchy of levels based upon degree of informational integration. Integrated Information Theory was developed by Giulio Tononi to describe the structure and content of integrated information generally, and has furnished researchers with a valuable mathematical tool that is applicable to the phenomenal symmetry space. The nodal network is a "neural network"—broadly defined in fields of biological and artificial intelligence as a network in which pathways become value weighted as the result of repetitive usage or by the encodement of trial-and-error success criteria. For example, value weighting in the nervous system may be attained by the addition of neuronal fibers to a pathway bundle, or, in response to contextual pressures, by the emission of neurotransmitters into widespread areas of the cortex via projections from subcortical value nuclei, such as dopamine from the ventral tegmentum to increase motivation, or serotonin from the caudal nuclei to prioritize emotion.

The integration of information through the nodal hierarchy begins when impulses, such as light rays or sound waves, impinge on sensory receptors in the eye or ear. Patterns contained within these impulses are mapped via the thalamus to first-level processing nodes in the primary cortices specialized for vision or audition. These processed patterns are then forwarded to second level nodes for unimodal association, where, for the sense of vision, specific colors, shapes, and motional attributes are identified and integrated. At levels three and four of the hierarchy, categorical information, such as object types and faces, or word-forms in the case of audition, is constructed. Nodes at levels five and six, the topmost levels, are transmodal, and specialized to interconnect and integrate multiple modalities.

As signals rise through bottom-up sensory channels of the nodal hierarchy, there is increasing two-way crossover cycling with dual top-down action channels specialized to carry out motor operations. Additionally, at the higher transmodal levels, other types of signaling cycles from non-sensory and non-motor nodes also function in the network, such as from nodes in limbic cortices which transmit signals concerning visceral, emotional, and motivational priorities and contextualization. Interconnection among the distributed transmodal nodes is facilitated by self-encoded pointer directories—as well as by frequency alignments and resonances to be discussed below—which produce the overall result of tying together coherent memories, thoughts, actions, and experiences. Note that integrated information can be stored and recalled subliminally, as well as consciously, via the cycles of the transmodal network, and that certain transmodal cycles may operate in the absence of any sensorimotor signaling.

Three transmodal nodes are noteworthy with regard to mental consciousness: the hippocampal complex, working memory, and the language processing system. The hippocampus is a three-layered cortical structure involved with the basic survival benefits afforded by spatial navigation, temporal orientation, and the temporary recording and retrieval of explicit (conscious) memories. The hippocampus—as the system has been termed when referring to the mammalian brain—is homologous to the three-layered reptilian archicortex, which lacks the developed six-layered neocortex and associated advanced cerebral capabilities found in mammals. The operational capacities of the hippocampus have proliferated as the mammalian brain has evolved, supplementing a large variety of cortical and subcortical functional centers. In particular, the hippocampus now supports the higher levels of the neocortical nodal hierarchy via the entorhinal cortex where the transition from three to six layers is made.

The short-term/working memory complex is located in the prefrontal cortex at the top of the nodal hierarchy. The prefrontal cortex is the latest neural region to have evolved in the human brain. It is a tightly knit group of ~15 nodes, some of which overlap other regions of the cortex. Short term/working memory nodes activate distributed representations stored in long-term memory cycles. Working memory is also central to executive-motor administration, where it operates through the top-down channels of the nodal network, screening the rise of signals at each level of the bottom-up channels in order to grant further ascension only to those signals that meet the current contextual requisites for contribution to higher representations of integrated information.

The Wernicke and Broca nodal areas operate together with other areas in the temporal and frontal lobes which are implicated in the comprehension and expression of language. The Wernicke-Broca nodes are part of a dual-pathway auditory system, running from the unimodal auditory cortex to the prefrontal cortex. This system is responsible for grammatical syntactation, storage of phonological (vocabulary) and semantic (meaning) lexicons in the long-term memory areas which intercept the dual-pathway streams, and phonological symbolization in conjunction with short-term/working memory. Lexical symbolization, being a transmodal functionality, generally activates non-auditory cycles as well; for example, when the word "cat" evokes into consciousness the visual image of the animal or an emotional feeling associated with it.

In addition to short-term/working memory and Broca's area, another major prefrontal region is the frontopolar cortex, which forms the most anterior portion of the prefrontal cortex. The frontopolar cortex is the largest region of neocortex in terms of uniform cellular composition, and, residing at the top of the bottom-up nodal hierarchy, provides the highest level of sensory, somatic, and abstractive informational integration; as well as being at the apex of top-down executive functionalities controlling intentional behavior, including attention, memory recall, cognition, and emotional-neurochemical modulation—along with their experiential correlates.

The "bottom-up/top-down" convention of naming the contra-directional corticocortical (within the cortex) pathways of the nodal hierarchy, while being fundamental causal vectors in the interface between the cortex and the rest of the central nervous system, must be qualified. These channels are densely and reciprocally interconnected, mutually modulating, project lateral feedback and feedforward cycles spanning multiple levels, and incorporate cycles from endogenous open-system process-structures which are themselves centers of emergent functionalities. The architecture of the nodal network is often better viewed as a complex of bidirectional cycles adjoining and unifying a constellation of biosemiotic convergence-divergence nodes. According to the "fourth law of thermodynamics" (Morowitz's Cycling Theorem), an open steady-state system (unchanging through a period of time) must entail cycles; hence the process-structure of the nodal network is properly understood to be an open-system set of holistically coupled and nested cycles.

* * *

In addition to its spatial organization, the brain operates in a symmetry space of temporal dimensionalities. We have mentioned that frequency alignments and resonances among distributed nodal structures are factors in the production of integrated information. The basis of the brain's temporal organization lies in the collaborative properties of its interneurons. Small circuits composed of interneurons and principal neurons send out messages about local firing patterns, via long-range principal neurons, to spatially-dispersed targets. These targets return messages reciprocatively, which induces the firing patterns of the local and distal structures to synchronize, or phase-lock, their oscillation frequencies, generating a network of spatiotemporally organized functional structures called a *neuronal group.* A neuronal group typically holds a synchronous pattern for hundreds of milliseconds, although a given pattern may be reconstituted and sustained by means of recursive cycling. In the case of informational patterns subserving the function of mental consciousness, recursive cycles are said to reenter the patterns, with the effect of endowing consciousness with a coherent temporal flow of *remembered presents.*

The synchronization frequency of a neuronal group is called its *brainwave* frequency. A brainwave frequency reflects the emergent global firing pattern produced by a neuronal group's constituent structures; although a neuronal group's collective brainwave frequency is generally different from the frequencies of the individual neurons that comprise it. A brainwave frequency is a network-level product of interneuron modulated microcircuits. Collective brainwave synchronization enhances the effectiveness of neuronal actions and responses because correlated signals impose a stronger impact on structures than those which are temporally disorganized; in-phase signals summate and amplify whereas randomly phase-shifted fluctuations cancel one another.

The generation of synchronicity and related temporal effects, such as resonance and phasal time locking, occurs across a broad range of brain locations and frequencies. For example, during periods of deep, dreamless sleep, the thalamocortical system oscillates in the *delta* frequency band [0.1-4 Hz]. This is the period of sleep when long-term memories are consolidated and encoded into the neocortex. *Theta* frequency brainwaves [4-8 Hz] supplement delta during deep sleep, and strong theta activity in the hippocampus is crucial to the construction of new explicit (conscious) memory cycles in the active waking state. Theta waves are also involved in selective attention, motor control, and language processing and integration. *Alpha* waves [8-13 Hz] have been correlated with visual activities of the occipital lobe, periods of relaxation, and top-down processes of working memory. *Beta* waves [13-30 Hz] are associated with a large variety of functions including the regulation of basic motor functions and top-down organization of the nodal network. In the frontal and parietal cortices, beta synchronization increases with working memory load and intensified attentional focus.

The fastest brainwave oscillations lie in the *gamma* bandwidth. The term gamma is typically employed to denote frequencies in the low-gamma bandwidth [30-70 Hz], shown to be involved with advanced neural functionalities in the upper levels of the nodal hierarchy, including the integration of multimodal information, cognitive processing, and working memory. Notably, ~40 Hz gamma brainwaves enfolding the thalamo-basal ganglia-corticocortical system have been correlated with various forms of conscious content, both in the waking and dreaming states. Higher-frequency gamma waves [70-90 Hz], together with alpha and beta, are involved in directing top-down attentional focus.

Interdependencies among brainwaves of different frequencies are a primary factor in the optimization of global operational synchrony, as indicated in the above cases of multiple frequencies affecting attentional focus. The interplay among brainwave frequencies engages both constructive and obstructive mechanisms based on the transmission characteristics of neurotransmitters. Neurotransmitters may be excitatory (e.g., glutamate, dopamine), tending to induce firing of a target neuron, or inhibitory (e.g., γ-aminobutyric acid, serotonin), tending to prevent firing. Principal cells are predominantly excitatory while interneurons are primarily inhibitory. The interplay of excitatory and inhibitory signaling regulates brainwave synchronies and their effects on neural operations. For instance, a beta synchrony may tend to unify a neuronal group but alpha synchrony may inhibit it, so that an energetically weaker but more salient alpha group may govern in lieu of beta; generally, slower rhythms provide contextual *temporal framing* for higher frequency, more specific, informational ensembles.

Experiments employing sophisticated neuroimaging techniques have correlated specific configurations of temporally framed neuronal groups with certain functions and informational types, including, but not limited to, objects of consciousness. The ubiquity of brainwave framing across conscious and non-conscious states has been substantiated through experiments of cognitive psychology: Certain experiments performed during waking and sleep states have identified brainwave framing for *neural correlates of consciousness*, while other experiments addressing non-conscious implicit memory and subliminal perception also indicate the involvement of brainwave framing. Spatiotemporal uniformity of neuronal grouping mechanisms is driven by the selective advantages of consistent, cross-platform holistic integration. The human microcosm integrates within and among all of its emergent platforms in order to implement its holism: from catalytic sets integrating chemical pathways, to organ systems integrating bodily functions, to neuronal complexes integrating abstract information. Consequently, at the highest level of integration in the brain, the neuronal groups and informational patterns empirically correlated with objects of mental consciousness have evolved so as to complement and enrich all other functionalities, thereby providing a tremendous selective advantage to the human microcosm.

The term "neural correlate of consciousness" has two connotations. The *full neural correlate of consciousness* is the spatiotemporal symmetry space that encompasses all of the indispensable process-structures which may combine to produce an object of consciousness. The collection of spatiotemporal structures corresponding to the full neural correlate of consciousness has been termed the *dynamic core*—the concepts of the dynamic core, neuronal group selection, the remembered present, and reentry are due to the late Nobel laureate Gerald Edelman. Originally, the dynamic core was conjectured to consist of the thalamocortical system, spatially, and 40 Hz brainwaves, temporally, but based on more recent evidence, the range of the dynamic core must be extended to include additional temporally-framed spatial constellations that dynamically morph through experienceable configurations. The term *content-specific neural correlate of consciousness* refers to two related physical structures: it is used to denote the minimal, indispensable process-structure of neurons active in the cortex during an individual instance of consciousness, or it may designate the information encoded in that process-structure. These two physical structures that characterize content-specific neural correlates of consciousness correspond to the experiencer and the experienced, respectively.

A group of experiments contrasting brain locations activated during periods of waking-state consciousness, and during sleep when the brain is disconnected from sensory and motor/task-oriented systems, has isolated neural correlates common to both states of consciousness. These tests demonstrate that primary subcortical structures, including the brainstem, basal ganglia, and thalamus, while critical to the support of various modes of conscious content in some circumstances, may be inactive in others, and therefore are not minimal, indispensable correlates of consciousness. With regard to the cortex, during periods of REM (Rapid Eye Movement) sleep, consciousness occurs in the form of dreams, while during deep sleep consciousness is absent. Combined, the results of these experiments indicate that (portions of) all objects of consciousness involve cortical activity in the so-called temporo-parietal-occipital "hot zone," generally with top-down contributions from various anterior regions.

The hippocampus is located in the temporal lobe region of the hot zone. Functionally, the hippocampus is a component of the medial temporal lobe memory system, which interacts with unimodal and multimodal regions distributed throughout the cortex to form, temporarily store, and then bind together whole, long-term, consciously recallable memories.

Another important complex located in the hot zone is the inferior parietal lobule situated at the junction of the temporal, parietal, and occipital lobes. This region is richly interconnected with visual, auditory, and somatosensory association areas, the frontal lobe, and other high-order assimilation areas. The finding that the temporo-parietal-occipital region of the cortex is implicated in all experiences of integrated sensory information marks this region as a home-base of ultrasensation, and thus of mental consciousness. However, while the more recently evolved frontal cortex is specialized for allocating intentional, attentional, and other executive functionalities, the qualia derived from these highly-developed top-down cycles are also integrated into objects of consciousness.

Qualia associated with bottom-up and top-down pathways are both descended from the primal underpinnings of the interneuronal domain's collective structure, its self-centered orientation, and its first-person vantage point. Early-appearing, but still extant mammals and birds are seen to be equipped with a neural substrate consistent with consciousness support: a phylogenetically conserved brainstem, diencephalic structures including the thalamus and hypothalamus, and a bidirectional nodal network capable of internal feedback and recurrent processing. The primitive lamprey, for example, is capable of selecting certain behavioral outputs while inhibiting others. The lamprey's top-down pathways, running from the cortex through the basal ganglia, are basically the same as those in the human brain, which regulate intentional behavior and consciousness processing. Further evidence illustrating the primitive origins of intentionality with respect to mental consciousness comes from experiments on canines in which all of the six-layered neocortex of the forebrain was surgically removed. These mammals retained the autonomous behaviors of integrated locomotion and orientation exhibited by amphibians, fish, and reptiles, which have no developed neocortex. In contrast, when both the neocortex and the phylogenically older three-layered allocortex (hippocampus, and the olfactory system and striatum) were removed from the canines, all intentional behavior was abolished.

Of utmost significance to the topic of intentional mental consciousness is a group of experiments in which brainwave anomalies peculiar to expert meditators were identified. A group of Buddhist monks, some at the behest of the Dalai Lama, participated in a cooperative study conducted at more than 20 universities over a 15 year period. Each monk had more than 10,000 hours of experience in meditation. The results of this project indicate that experienced meditators, compared to norm groups, possess extraordinary abilities to control their bodies and brains, and that the structural anatomies of their brains actually transfigure over time. Long-term effects of meditation on the brain were also found to depend upon the type of meditation practiced. In these experiments, three traditional meditation techniques were analyzed: one-pointedness, mindfulness meditation, and compassion meditation.

The meditative practice of *one-pointedness* is the holding of a single object of consciousness in focus for an extended period of time. One-pointedness was the original technique of yogic meditation. To achieve one-pointedness requires that all unrelated cycles, including extraneous sensory signals, visceral feelings, breathing, recursive thinking, etc., be intentionally inhibited from integration into an object of consciousness. In the contrasting practice of *mindfulness meditation,* informational cycles are intentionally welcomed into a continuously changing flow of consciousness, but the particular cycles associating excessive emotion and worldly attachments into consciousness are reflectively inhibited. Over time, mindfulness meditation develops one's ability to cope with distressing thoughts and traumatic events. *Compassion meditation* requires the continued effort to cultivate awareness of the suffering of others, and to intentionally experience sincere concern and empathy for them. These three meditative practices of choosing and fashioning one's own phenomenal symmetry space through the prolonged intentional regulation of their conscious experiences have proven to be of great value for those who have employed them, both personally and in societal relationships.

Expert meditators are able to manipulate the locations and amplitudes of electromagnetic activities throughout their nervous systems. Through neurofeedback, they can take intentional control over their autonomic nervous systems and bodily functions in a manner distinct from those evincible by non-adepts, as exhibited by the physical feats they perform and their conspicuous purity of being. The experiments indicate that yogis can intentionally control electrical activities throughout their nervous systems by generating ultra-high-amplitude electrical spikes in the palms of their hands through techniques called *mudras*. In the brain, one-pointedness and mindfulness meditation have been correlated with sustained high-amplitude gamma-band oscillations and phase synchrony in the range between 25 and 42 Hz, particularly in the frontal lobe, as well as increased power levels and spatial distribution of alpha oscillations. After years of practice, high-amplitude theta-band rhythms come to spread over much of the brain.

The meditation experiments also demonstrate that long-term practice causes regions of the brain to change in size, with specialized meditative techniques corresponding to size alterations of specific neuroanatomical regions. In the practice of mindfulness meditation, for example, persistent inhibition of emotional attachments results in decreased amygdalar volume. In general, both the prefrontal cortex, involved with intention, attention, and executive functions, and the insular cortex, folded deep within the hot zone and involved with consciousness of homeostatic information, emotion, compassion, and self-awareness, increase in thickness for meditators as compared to control groups. Other alterations to brain structure resulting from long-term meditation have been shown to include changes in neural pathway densities and the remapping of interfunctional transmodal networks.

The neuroplasticity of the brain allows meditators not only to intentionally design and mold their own physical *vehicles* of consciousness, but also to customize and experience altered *states* of consciousness.

According to both the rishis of scriptural lore, as well as the yogic masters participating in experiments of today, the most profound effects that may be realized through the dedicated praxis of meditation pertain to the nature of phenomenal experientiality itself. Yogis universally report that their experiences obtain novel characteristics, and as their minds and bodies progressively transform, their sense of subjectivity wanes. Experiencer and experienced merge into oneness. As the anthropic pathway has wended its way through the evolving macrocosmic universe, consciousness emerged as a powerful potentiality of life, destined to attain the wondrous heights of experiential reality in the phenomenal symmetry space of the human brain, where it renders unto each of us mastery over our very own microcosmic universe.

* * *

Part 1
The Evolution and Nature of Consciousness

1

Symmetry

In his masterpiece, *Elements,* Euclid used the word *summetria* (Gr. with measure) to denote commensurability among the parts of a geometric whole. Other early thinkers added philosophical connotations to this geometrical meaning of symmetry. Plato, for instance, in *Timaeus,* exalted the symmetries of the five regular polyhedra (the tetrahedron, cube, octahedron, icosahedron, and dodecahedron) as the underlying bases of the natural elements (respectively: fire, earth, air, water, and universe). The symmetries of geometrical figures are today defined in terms of their invariances under specified groups of rotations and reflections. When a cube is viewed perpendicularly to one of its faces, for example, it looks like a square, and particular 90° rotations of the cube from that position also look like a square. This *group* of rotational *transformations* leaves the appearance of the cube *invariant.* This sense of geometrical symmetry group transformations was later extended to include the interchange of congruent parts of a whole; hence, exchanging square faces of a cube leaves the appearance of the cube invariant.

By the seventeenth century, group symmetry theory was being applied to such abstract invariant wholes as the mathematical form of a physical law,[1] whose transformations navigate all of the possible *representations* of the law's symmetry group. The invariant form of a physical law often contains a combination of different types of symmetry, and an intriguing panoply of devices is employed by nature to uphold them. In Chapter 13 we will review the relationships and overlaps among exact, gauge, global or local, continuous or discrete, and integrated or broken symmetries. All of these expressions of symmetry pertain to the nature and evolution of consciousness.

Symmetry involves both diversification and unification. When diverse entities possess a common property, they are said to be symmetrical, as circles of diverse diameters are symmetrical because they each possess the geometrical property of roundness. In the particular case where a group of symmetrical entities is complete, i.e., the group encompasses all possible representations of a given property, it constitutes a unifying *symmetry space*. Thus, all possible shapes possessing the property of roundness, regardless of their diameters, are unified in the symmetry space of circles. An important class of symmetry spaces applies to the states that a dynamic system may come to occupy. In this context, the unifying property of the symmetry is the specific transformation through which the system evolves. The standard model of particle physics, for instance, is formulated in terms of symmetry transformations.

Another type of transformational symmetry, recognized to be operating across multiple provinces of nature, is the *conformal symmetry*—which unifies across size scales. Two examples illustrate the extent and influence of conformal symmetries in nature: In the macrocosm, a conformal symmetry transformation regulates the big bang transition between cycles of the universe as it passes from the cold vastness at the close of one cosmic cycle into the hot tiny seed initiating the next cycle. In the microcosm, conformal symmetries unify the fractal (self-similar) intra-dimensional interaction networks prevailing at all levels of living system organization. In the cerebral cortex of the human brain, at the nexus of its fractal neural network, scale-free dynamics seamlessly knit the global workspace of our unitary consciousness.

It is useful in the discussion of symmetry transformations to borrow a concept from physics called the *Lagrangian*—a mathematical expression of symmetry that encapsulates a system's dynamical properties. As a system changes from one state to another, the *form* of its Lagrangian remains invariant—that is, the symmetry specified by the form of the system's Lagrangian is a law of nature. The symmetry-based Lagrangian formalism is very powerful. For example, by first postulating certain symmetries as being fundamental properties of nature, Albert Einstein was able to derive the algebraic equations that quantify his special and general theories of relativity. Before Einstein, physical theories had been deduced from statistical consistencies and measured invariances, such as the total energy or the electrical charge of a system. Einstein's great philosophical breakthrough was to reverse that trend and use symmetry principles as the focal point of physical law from which the equations follow. Since Einstein, numerous conjectures about the symmetries of nature have guided physicists to important advances in both the macroscopic and the microscopic domains.

Throughout the course of cosmic evolution, the symmetry space of the universe admits innumerable representations and transformations. In the particular transformation of the universe called the big bang, even cycles of the cosmos are correlated in symmetry, but description of that transformation requires terms transcendent to the familiar notions of space, time, and matter. According to modern theories of big bang cosmogony, arrows of time (toward the past and future), physical entities and their environmental dimensionalities, strong and weak coupling strengths, and large and small distance scales (the M-Theory dualities) were once unified in a parent *supersymmetry* which can only be described in terms of abstract mathematical moduli. The idea of a big bang transformation being an event on a unidirectional timeline must yield to the holographic picture, where macroscopic dimensionalities, such as time, are subsumed in a higher symmetry. The transformational nature of the universe depicts the big bang as a nonlocal steady state fount of manifestation issuing forth from a transcendent supersymmetric source (hence, the steady state theory posits a process of continuous creationism as an alternative to the big bang). As we shall see later in the discussion of emergent monism, steady state characteristics are fundamental to the essence of the universe. Nevertheless, from our vantage point on the cosmic landscape, numerous observations and deductions coming from physics, biology, and other scientific disciplines enable identification of a consistent symmetry-based paradigm that extends from the big bang to the present day.

In the past century, an extensive assortment of scientific theories have been formulated in terms of symmetry principles. As a result, it is now generally accepted that symmetry is a fundamental feature of reality. However, symmetries are unstable. They break and rebuild in a manner which causes the manifest world to emerge. As first observed by Pierre Curie in 1894 in his study of crystal structures:

C'est la dissymetrie qui cree le phenomena.
"The phenomena break the symmetries of laws."

— Pierre Curie, 1894. [2]

Manifestation occurs only when symmetries are broken. Thus, the direction in which a crystal actually grows, breaks the symmetry of the many alternatives in which it could possibly have grown. In addition to symmetries being broken locally by actualized circumstances, as with the crystal, symmetries are broken globally across the universe. The detection of the Higgs boson in 2012, by means of the Large Hadron Collider in Geneva, completed the verified list of "standard model" particles that had been previously predicted on the basis of symmetry breaking principles. By reaching temperatures in the collider comparable to those that existed in the newborn universe, the threshold of electroweak symmetry breaking through the Higgs mechanism was empirically duplicated. When the Higgs mechanism originally occurred immediately following the big bang, it had two profound effects upon the universe: It bestowed mass to particles, and it "Let there be light." The breaking of the electromagnetic interaction from electroweak symmetry liberated and energized the photon field upon which organic life would ultimately arise.

Cosmological symmetry breaking has had far-reaching consequences. Due to an inherent instability in the supersymmetry of the big bang universe (the tachyon[3]), a progression of *spontaneous symmetry breaking* transitions precipitated four residual sub-symmetries. These so-called "gauge symmetries" correspond to the four known force fields of nature; namely, the gravitational, strong nuclear, weak nuclear, and electro-magnetic fields (it is the weak nuclear and electromagnetic fields that were re-symmetrized into the electroweak field in the Large Hadron Collider). Gauge fields are responsible for all interactions among physical objects in the universe. As gauge symmetry is the basis for the theories of special and general relativity, quantum field theory, and string theory, it has been acclaimed, "the cornerstone of modern physics." [4]

* * *

2
Complexity and Holism

Although the supersymmetry of the universe broke into diversity in the big bang, the universe remained holistic at its fundamental quantum level, with all of its parts causally interdependent and imperatively unified. Elementary particles, such as quarks and electrons, which materialized through symmetry breaking, are actually sub-wholes within (representations of) the symmetry space of the holistic universe. In the microscopic world, where quantum physics holds sway, an elementary particle cannot be isolated and treated independently of its environment; that is, without inviting paradoxes such as those that plagued the early attempts to explain the strange particle behaviors first detected in the 1920s. These paradoxes were subsequently resolved with the *ontological interpretation* of quantum theory put forward by David Bohm in 1952, which, he said, "emphasizes the wholeness signified by quantum laws."

While sub-whole particles and their composite structures certainly do express the emergent levels of classical behavior with which we are familiar in our everyday affairs, they are also inextricably connected to the indivisible macrocosmic network. Sub-wholes and their macrocosmic basis together comprise the symmetry space of the universe, whose very holism is causal. Holism is an active, unifying, and driving mandate in the holomovement of the universe. (The term holism will be used in two contexts: 1) in physics, where holism refers to a macrocosmic property, signifying the interconnectedness that unifies the entire universe, 2) in the life sciences, where holism is a property of the microcosm, denoting the causal property of certain complex systems that emerges upon the closure of interdependent processes.)

A primary concern of big bang cosmogony is the implication that the universe must have initially held the potentiality for the generation of order, as it is seen to exist throughout the universe today. This seems to conflict with the observation that the universe, as a whole, has been evolving toward a state of disorder since the big bang, the statement of which is the second law of thermodynamics. In other words, the universe is expanding and settling into a spread-out state of disordered, high-entropy, thermodynamic equilibrium, akin to a gas that has escaped from a container and expanded into a diffuse and unorganized state. Initially, the distribution of particles across the universe was extremely uniform (of maximum entropy) and therefore the effects of gravity were cancelled-out in every direction in the homogeneous medium. Then, random fluctuations in particle density induced the gravitational field to activate degrees of freedom in the macrocosmic symmetry space. This made dimensional entropic "room" for the formation of orderly structures and complexity, despite the overall inexorable drift toward thermal disorder.

> *What was extraordinarily special about the big bang was actually its great uniformity…the gravitational degrees of freedom providing a huge reservoir of entropy.*
>
> — Roger Penrose [5]

Complexity, which is the measure of interdependencies within a system, requires the combination of order and disorder. Alone, neither ordered structures, such as crystals, nor disordered media, such as gases, can facilitate a complex network of interdependencies among their parts. The interface between order and disorder is an extremum that evolves toward an attractor of complexity, traversing the anthropic pathway, where intricate structures and complex process-systems may evolve.

Representation of the laws of physics in Lagrangian form captures a concept of symmetry in a manner that can be applied to complex open systems, including living microcosms (variable amounts of matter, energy, and information pass continuously through an open system, whereas in an isolated-system they remain conserved and contained; in physics, the Lagrangian specifies the rules governing the inner-dynamics of conserved systems only). This is our prime motivation for delving into the foundations and ramifications of symmetry. The underlying principles of symmetry which first dictated the physics of the early universe are also dominant factors in the evolution and dynamics of complex systems, including vehicles of consciousness. In the case of an organic open-system microcosm, it is the envariance (environmentally assisted invariance) of its Lagrangian that determines the transformations which may be actualized within its symmetry space (defined by its included holistic set of interdependent process cycles). Intra-biotic transformations that take place within a microcosmic symmetry space are extremely complex, entailing factors such as evolutionary adaptation of the envariant Lagrangian itself, contextual selection of optimal internal process networks, and the myriad varieties of matter, energy, and information passing from the environment through the open system. Each of these factors signifies the intimate, reciprocative, and unifying relationship that exists between an open-system microcosm and its open-system macrocosmic counterpart.

In recent years it has been established that the unidirectional arrow of time evoked by cosmological symmetry breaking, projecting from the big bang to the present, is a systemic factor in the integration of complex-system symmetry spaces. The arrow of time is deeply associated with the second law of thermodynamics, which mandates that the total entropy (the tendency for a closed system to evolve toward disordered, unstructured, equilibrium) of the universe is increasing. Today, we realize that entropy is also implicated in the ordering of living, nonequilibrium open systems: [6]

We can no longer associate the arrow of time only with an increase of disorder. The arrow of time is a source of order. Recent developments in nonequilibrium physics and chemistry point in the opposite direction. The physics of non-equilibrium processes describes the effects of unidirectional time and gives fresh meaning to the term irreversibility. We now know that irreversibility leads to the essential constructive role of the arrow of time. Without this new coherence due to irreversible, nonequilibrium processes, life on earth would be impossible to envision.

— Ilya Prigogine [7]

* * *

3

Selection and Self Organization

The term "natural selection" was coined by Charles Darwin in his book, *On the Origin of Species.*[8] He chose this term to contrast with "artificial selection," which refers to the exclusive propagation of plants and animals possessing certain desirable traits such as large size or pleasant taste. Artificial selection involves the intended purpose of human beings, whose inventions have affected the evolution of life. Other examples of purposeful (teleologic) invention by human beings that have resulted in significant benefit include genetic engineering and the design of vaccines to control, or, as in the case of smallpox, even eradicate diseases.

In the process of natural selection, as opposed to artificial selection, microcosms evolve to fitness through unintended trial-and-error fine-tuning by nature.

> *Nature is a tinkerer and not an inventor...It is not biochemical novelties that generated diversification of organisms. In all likelihood, it worked the other way around. It is the selective pressure resulting from ecological niches that led to biochemical adjustments and changes in molecular types.*
>
> — Francois Jacob [9]

This quote by Francois Jacob refers specifically to the kind of phenomena observed by Darwin during his voyage aboard HMS Beagle in the 1830s, where he observed that the finches inhabiting the Galápagos Islands included 13 closely related species, differing primarily in the shape of their beaks, in correspondence with the shape of the nuts they favored. Accordingly, natural selection is usually construed in terms of macrocosmic environmental pressures, such as the availability or type of resources guiding the successful adaptations of dependent microcosmic organisms. However, adaptations first emerge as random events within a microcosm which might happen to be advantageous in its interface with the macrocosm. This interrelationship between a microcosm's adaptive causality and the macrocosm's selective causality leads one to consider the ubiquity of the self-perpetuating processes of evolution as purposive (teleonomically), even though factors of randomness are involved.

Since Darwin, the underlying principles of selection have been found to apply in many domains, including physics (quantum Darwinism[10]), biochemistry (complex adaptive systems), genetics (natural selection), embryological and cognitive neuroscience (neural Darwinism[11]), and beyond, to the realm of the natural sciences, to economics and sociology (universal Darwinism) where, for example, evolving memes replicate and adapt in their social milieu. Each of these forms of selection involves an envariant symmetry relationship in which open-system processes adapt with and mold the very environment that selects them. In the Gaia philosophy, this notion of causal reciprocity applies to the relationship between macrocosm and microcosm as it concerns life on earth. For example, while the biosphere promotes adaptations in plants that optimize their homeostasis, the plants, in turn, alter and regulate their biosphere through photosynthesis and respiration.

In quantum Darwinism, the fitness for a particular species of particle to materialize from the field of quantum potentiality and survive as a stable entity is dependent upon the state of the environment being propagated by all other surviving and interacting particles. This interdependency between whole and sub-whole, first emerging in quantum interactions, initializes the reciprocative process of quantum Darwinism. At this basal quantum level, the evolutionary paradigm of selection is forged into the underlying envariant symmetry of the cosmos.[12]

A particle is a stable sub-whole within the holistic universe with which it is always conjoined in a reciprocative, intra-causal relationship. The fittest, most stable kind of environmental sub-wholes have been shown to be those for which the symmetry Lagrangian of the total-system is minimized. The conjoined symmetry space of microstructures (the evolutionary forerunners of complex microcosms) and the macrocosm is then said to be diagonalized, or decohered. Through this process of quantum Darwinism, the selection of particle pointer states gives birth to the building blocks of classical physics, from which our world of everyday objects and interactivities is constructed.

The pointer states selected through the process of quantum Darwinism in the early universe must have been endowed with precisely valued properties, or the potential to assemble those properties, which made possible the subsequent evolution of complex systems able to support conscious life. What was the pre-existing cause or reason for those properties to have had the improbable coincidence of values and intricate interdependencies required for complexity to ascend? This states the notorious problem of fine-tuning, which remains a contentious subject. The cosmological anthropic principle postulates an infinitude of parallel universes, comprising a "multiverse," such that one of these universes, ours, would likely possess the fine-tuned (anthropic) properties that we observe. This solution to the fine-tuning problem is quite extravagant in postulating an infinite number of universes, of which we can never know, because there is a difficult problem concerning our own universe.

A more parsimonious solution to the fine-tuning problem is to extend the principle of least action, which applies throughout our own universe. The principle of least action states that the mathematical extremum of the Lagrangian functional (df = 0) specifies the trajectory of representations in a symmetry space that actually becomes manifest. In the least action approach, the (ultra-dimensional) multiverse of randomly parametrized universes is replaced by a single, holistic, quantum-entangled (ultra-dimensional) symmetry space characterized by a Lagrangian comprised of deeply-nested, Darwinian-envariant spacetime cycles. This Lagrangian possesses an extremum between order and disorder on the "edge of chaos,"[13] whereupon evolution proceeds toward an anthropic attractor of complexity. Regardless of the reason our universe is fine-tuned—multiversal chance or universal attractor—the necessary and sufficient pointer states that have been selected in the quantum regime are manifestly established as the foundation upon which life and complex vehicles of consciousness have ascended.

The reciprocative relationship connecting macrocosm and microcosm underlies the propensity for certain systems to self-organize global properties. Two simple laboratory experiments demonstrate systems that spontaneously self-organize global properties that cannot be accounted for on the basis of local interactions: In the convection dish experiment, liquid chemical mixtures, when heated from below in a shallow dish, self-organize to form a regular *spatial* array of hexagonal convection cells; in the chemical reactor called a Brusselator, a mixture of chemicals displays *temporal* self-organization as it cycles through a spectrum of colors. In order to describe self-organization in physical systems, the methods of complex attractor dynamics have been developed. This approach enables one to infer the plausible and intuitive process which is conducive to electrochemical self-organization, and which promotes the evolution of dissipative structures (open systems) and living microcosms. These nonlinear dynamics taking place in substrates of limited resources are fundamental to natural selection.

From the instant of the big bang, 13.8 billion years ago, to the present, complex systems have been evolving in accordance with an anthropic principle. This mode of evolution has proceeded along a very distinct pathway, through an envariant symmetry space where the macrocosm and microcosm have reciprocated to optimize the selection of adaptive, inventive, self-organizing systems. The anthropic pathway seeks out an attractor of complexity, finding and fortifying conditions optimal for the macrocosm and the microcosm—environment and living being—to coevolve. The foundations upon which selection-invention have been implemented, have themselves evolved: advancing from the symmetrical underpinnings of quantum Darwinism, which gave rise to the elementary particles, to the chemical substrates of holistically complex autocatalytic sets, which expanded through natural selection into the physiologies sustaining and empowering living organisms, to the neural networks of the brain, where objects of mental consciousness are integrated through neural Darwinism for subjective experience in the phenomenal world of the human mind.

* * *

4

The Origin and Ascent of Life

The "simple" total-symmetry constraint governing envariant quantum Darwinism, which suffices for the selection of elementary particles, has evolved in life systems to become an elaborate interrelationship that maintains balance between selecting macrocosms and inventing microcosms, and vice versa. This reciprocative cycle nurtures the ever-increasing fitness of open-system microcosms by integrating inter-dependent process pathways into complex symmetry spaces, be they chemical autocatalytic sets, bulk material transport systems, or nervous systems. The primary feature that distinguishes the varieties of integrated microcosmic symmetry spaces is the degree of their spatial and temporal infoldment.

Representations of integrated information in the symmetry space of the human brain have been correlated with sophisticated conscious events. Manifestations of advanced consciousness are enabled by the brain's extreme complexity, the result of the infoldment of its spatially-packed, temporally-synchronized neural configuration. Not all conscious beings possess a highly infolded centralized brain, however. The degree of infoldment that characterizes the integrated symmetry space of a living system varies widely in nature. In this section and the next, we examine the continuous evolution of integrated symmetry spaces, beginning with deep-sea vent chambers, continuing through the formation of membranous unicellular and multicellular organisms, and culminating with the human brain. We shall see that the evolution of vehicles of consciousness has proceeded continuously, without missing links, and that a key selective force has been the coming together—the infoldment—of integrated symmetry spaces.

Immediately following the big bang, local gauge symmetry interactions induced elementary particles such as quarks to bond into stable nucleons (protons and neutrons), and these into the light atomic elements of hydrogen, helium, and lithium. Globally, the long-range gravitational force attracted these light elements to form nebulous galactic clouds, followed by distinct stars. Within the stars, gravitational convergence continued until mounting pressure so intensified that runaway nuclear reactions were triggered, fusing together and exploding the full complement of atomic elements into the interstellar medium. There, under the unremitting pull of gravity, these elements consolidated to form molecular clouds, new populations of stars, and metal-rich planets (in cosmology, the terms "metal" and "metallicity" denote all of the atomic elements except hydrogen and helium). Earth is a metal-rich planet, with all of the elements necessary to support the organic chemistry of living systems.

The Hadean eon was the first geologic period on earth, beginning 4.6 billion years ago and continuing on to the advent of the Archean eon, 0.6 billion years later. Conditions in the Hadean eon were torrid, hellish, yet life emerged. After a period of cooling, a liquid water ocean saturated with carbon dioxide condensed upon the surface of the globe. The magma in the hot molten core below was in a violent state, spewing hydrogen, other gases, and acidic minerals into the surrounding ocean waters through the few volcanos that rose above sea level at that time, and through solidifying underwater structures called "deep-sea vents," or "black smokers."

Black smokers were first discovered along the Galapagos Rift in 1977. Since then, other active black smokers have been found, including the Juan de Fuca Ridge off the coast of Oregon. Precipitation of the effluents from these structures onto the ocean floor has had two notable effects: At the base of the black smokers, highly concentrated mineral deposits have supplied the nutrients that fueled the rise of unique chemotropicbased food chains, supporting such exotic animals as giant clams and tube worms, translucent fish, shrimp with eyes on their backs, and glowing organisms able to light a world more than a mile deep in the abyss. Secondly, the effluent build-up created a sea floor of spreading and colliding tectonic sheets. Through the geologic ages, some of these sheets have slid under others, pushing them upward to form the great mountain ranges of our present eon, including the Himalayas and Alps.

In addition to the violent black smokers and volcanos sprouting up in the Hadean eon, the earth's churning mantle gave rise to cooler, less turbulent underwater structures, where life was born. In this case, metal-rich minerals *reacted* with the seawater to form porous structures called "alkaline deep-sea vents." Alkaline deep-sea vents were first discovered in the year 2000 in an underwater mountain range called "The Lost City," so named for its location in the Atlantis massif, the hypothetical site of the fabled civilization of Atlantis. Alkaline vent structures are riddled with interconnecting cell-sized chambers. In the Hadean environment, these chambers were endowed with a metabolism derived naturally from the energy available through the exothermic reaction between the hydrogen coming from inside the earth and the carbon dioxide in the ocean. This continuous source of energy facilitated the bonding and evolution of increasingly larger and diverse molecules within the chambers. The walls of the chambers became lined with catalytic metallic solids (e.g., iron, manganese, zinc, molybdenum, nickel, tungsten, and their sulfides) which promoted the formation of reaction pathway interdependencies and catalytic sets needed for the advancement of complexity.

Temperature variations in the vent structures established a vertical reactivity gradation for inter-chamber catalytic chemical pathways. In the hotter regions near the bottom of the structures, monomer synthesis was energized, while in the cooler, more tranquil regions near the top, polymerization of fragile molecules was possible. Along the temperature-graded pathways that flowed up through the labyrinth of interconnected chambers, cellular evolution unfolded through the complexifying stages of prebiotic chemistry, monomer synthesis, polymeric nucleotide/peptide synthesis, the RNA-protein world, and lipid biosynthesis.

The high acidity of the seawater outside a vent chamber in combination with the high alkalinity of the chemicals flowing into the chamber from below created an electrochemical potential across the chamber walls. The effect was a proton-motive force that provided manageable energetics for the reactor-chamber's metabolism. Chemically, the energy potential across a chamber wall was tapped through the development of precursor ATPase, a catalyst that produces ATP (adenosine triphosphate), the energy currency of all microcosmic systems living today.[14] The cooler chambers toward the top of the vent structures, wherein polymerization was thermodynamically feasible, were also farther from the strong ATP-inducing proton motive force of the hydrothermal fluids, so that catalytic sets able to regenerate the potential gradients, independent of direct hydrogen contact, could succeed and evolve in these regions. These processes eventually supplanted the primitive, passive metabolism of the chambers altogether.

Hydrophobic lipid substances developed naturally from the chemical ingredients flowing into the vent chambers. These oily lipids tended to coat the interior of the chamber walls, forming inorganic membranes that separated sites of oxidation from sites of reduction, located on opposite sides of the membrane. This established an electromotive gradient across the permeable membrane, which was harnessed for the transport of chemicals, such as quinones, into the space enveloped by the membrane, there increasing metabolic efficiency and fortifying the catalytic processes of self-organization. Eventually, autocatalytic chemical pathways self-organized, becoming capable of biosynthesizing an organic lipid-based membrane specialized for specific environmental interfaces. The lipid membrane superseded the protective shielding previously provided by its porous calcium carbonate encasement, equipping the open-system microcosm with an internally managed self-containment mechanism which was metabolized by an adapted ATP-based Krebs cycle. With the holistic autocatalytic microcosm now enclosed within a semi-permeable membrane, the self-sufficient *organic cell* was empowered to depart the vent structure and begin its marvelous odyssey into the world at large.

Within the organic cell, catalytic reaction cycles continued to proliferate and accelerate the production of various molecules; see Figure 4-1.

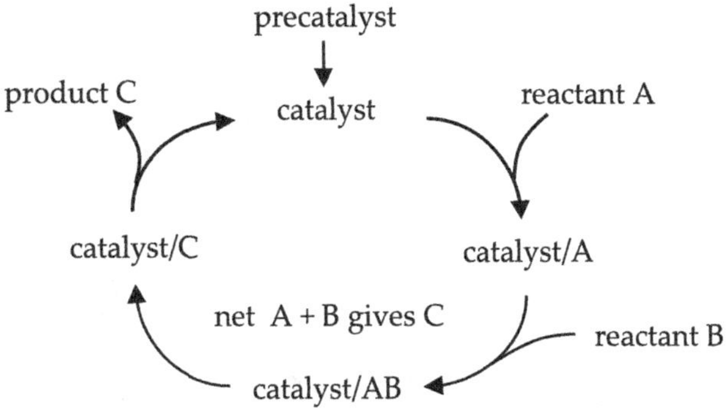

A catalytic cycle is a repeating sequence of chemical reactions that, under the influence of catalysts, increases the rate of production of a product. The catalytic cycle begins with the electrochemical binding of reactant A to a catalyst, yielding a compound (catalyst/A), which binds to reactant B (catalyst/AB), which then dissociates into the required product C = (A+B) plus the original catalyst which is now available for another cycle.

Figure 4-1 *Catalytic Cycle*

Some of these molecules were, in turn, capable of catalyzing even larger and more diverse compounds. As a result, new generations of reagents and catalysts affected the existing chemical processes via feedback and feedforward pathways, such that stable, self-sustaining, self-modulating autocatalytic sets were established.

An *autocatalytic set*[15] is a collection of molecules that catalyze each other's production. An autocatalytic set may comprise a simple, closed-loop hypercycle (a sequence of reactions in which each molecule in the cycle exclusively catalyzes the production of its successor, such that the last reaction catalyzes the first), or it may comprise an intricate network of coupled and nested catalytic pathways. Autocatalytic sets act collectively, forming a self-sustaining, self-organized whole that can reconstitute itself after a large portion of the set has been removed. A simple autocatalytic set is depicted in Figure 4-2.[16]

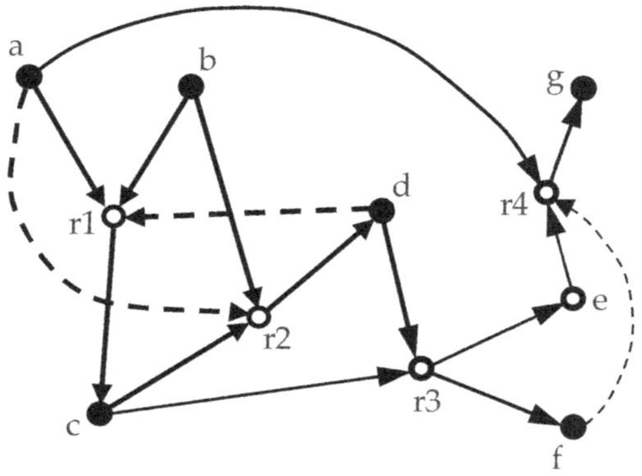

Autocatalytic set with seven molecule types {a, b, c, d, e, f, g - solid nodes}, and four reactions {r_1, r_2, r_3, r_4 - open nodes}.

The food set is {a, b}. Reactants and products are represented by solid lines, catalysis by dashed lines, and flow capacities by line thickness.

Figure 4-2 *Autocatalytic Set*

Autocatalytic sets require replenishment of resources passing through their open-system architectures and must therefore maintain effective relationships with their host environment. Microstructures, such as protons, atoms, and small molecules, which furnish the ingredients for autocatalytic sets, are only the evolutionary precursors to holistic microcosms. They are not open systems (above the quantum Darwinian level), as they lack an internal, environmentally responsive, selectively adaptive repertoire of autocatalytic process cycles. There is a fundamental connection between open systems—of which all living beings are examples—and the cyclic processes within them. A theorem proven by Harold Morowitz states:

> *The flow of energy through an open steady state system will lead to at least one cycle within that system.* [17]

Initially, the process pathways constituting the autocatalytic sets in vent chambers were dispersed throughout the chemical mixtures. In terms of relative concentrations, the formation and dissolution rates of these chemicals self-modulated autocatalytically so as to sustain a natural balance with respect to the available, and often limited, environmental resources. Continuous adaptations of the microcosmic pathway cycles to the changing environment, through their mutual catalytic support and refinement (in particular, the formation of autocatalytic cores, i.e., autocatalytic cycles competing for resources[18]) ensured the stability and evolvability—the survival of the fittest—of interaction networks. This mechanism facilitated the natural selection of the most temporally responsive (temporally infolded) microcosmic pathway sets and communication webs, as well as the most efficient, spatially localized (spatially infolded) functional structures. Complementarity of temporal and spatial infoldment within the autocatalytic symmetry space expedited the ascent of microcosmic complexity and fitness.

Autocatalytic sets promoted the synthesis of nucleotides and peptides. These particular molecules orchestrated a reciprocative interaction cycle that laid the foundation for the evolution of extremely complex living microcosms. Nucleotides and peptides co-developed into RNA and protein chains, respectively. Pre-RNA nucleotides were able to self-polymerize into long strands by means of the sequential catalyzation of selectively advantageous nucleotides onto their malleable allosteric end. RNA molecules were unique: in conjunction with other abiogenic reactants present in the substrate, they could catalyze their own production, or self-replicate. That is, RNA molecules could reproduce, in one shot, all of the instructions encoded in its nucleotide sequence necessary to catalyze the full set of pathways that had evolved over the long course of evolution. RNA could also interpret and translate various chemical structures so as to produce, or *express,* catalysts necessary to synthesize more RNA (genotype), as well as a population of proteins (phenotype) complementary to messages received by the RNA complex.

The primary structure of an RNA molecule resembles half of a ladder, consisting of a sequence of nucleotide "rungs" projecting from a sugar-phosphate backbone. The informational content of the ordered sequence of nucleotides in an RNA molecule specifies the *genetic code* of instructions that determines how the RNA molecule catalytically modulates chemical process networks. Some sequences of nucleotides along an RNA strand can catalyze reactions involving many types of chemicals, including amino acids and other RNA species, whose subsequent interactions effect the *genetic expression* of the RNA molecule. Enabled by catalysis of its own sugars, bases, and phosphates, RNA (ribonucleic acid) next adapted to form the DNA molecule (deoxyribonucleic acid) to acquire the selective advantage of structural durability. DNA is basically two strands of RNA connected intermittently by complementary nucleotide pairs, forming a twisted, ladder-like, double helix structure called the "spiral staircase." Strong and resilient DNA molecules were able to grow to extreme lengths, billions of nucleotide pairs long, and infold more pathway catalysts into a single-molecule information processing center, dedicated to the management of the microcosm's holistic integrity. The physically stable DNA molecule is effectively the infolded core of a giant intra-modulating catalytic web of coupled and nested process cycles. It is a dynamic player that can change its shape and chemical reactivities in response to environmental context. In return, DNA is the genetic arbiter of the functional cycles that constitute the phenotypic symmetry space of the microcosm.

The functional process cycles peripheral to the orchestrating DNA hub also evolved toward increased infoldment and complexity. Functional cycles consolidated into compact complexes called "organelles." Over generations, the organelles adapted into highly efficient processing centers; for example, the protein-synthesizing ribosome and the lipid-synthesizing endoplasmic reticulum. Lipids formed into semi-permeable membranes embedded with specialized protein molecules covering the cell wall, blocking entry of unfit molecules while permitting flow-through of necessary resources.

The sets of interdependent process cycles within the self-contained microcosm constituted an "RNA-protein world" that was *collectively* responsive to environmental variations, assuring the integration of those cycles most fit to respond and adapt to environmental change. These synergistic networks had garnered the key characteristics of holistic biotic systems; they possessed self-proficiencies: they were self-modulating and self-replicating autocatalytic sets, self-organizing complex adaptive systems, self-renewing autopoietic process-structures, and self-referential discriminators of internal versus environmental signaling. They were holistic microcosms—they were living organisms.

The type, quantity, reaction strength, compatibility with internal cycles, and other properties of the chemicals entering into the open-system microcosms, conveyed biosemiotic signs (signal structures) that were complementary to the interpretant structure (the process-structure of the resident holistic microcosm). In effect, the microcosm was a contextually flexible interpretant, integrating signs so as to instantiate and act upon pertinent semantic information. The capability to translate signs, such as nutrient levels or the presence of a dangerous object, into meaningful information relative to the interpretant process-structure, heightened the fitness of the microcosm to respond to environmental pressures and opportunities, and became a prime driver in the evolution of complexity systematics.

Once deep-sea vent microcosms had adapted for lipid membranes and departed from their chambers, two primary types of cells evolved. Each had traits that assured fitness in a changing environment. One of the cell types was bacteria. These were simple organisms whose success was derived from their prodigious ability to replicate. The other type was archaea. Archaea depended on complexity for their fitness. They were master inventors, readily adaptable to environmental circumstances. Gene sequence studies indicate that significant increases in organic complexity and symmetry space integration occurred two billion years ago through a process called *endosymbiosis*—the engulfment of one type of cell or organelle by another, with the merger of their functionalities. In the so-called "fateful encounter," [19] an archaeal cell engulfed a bacterial cell, resulting in the formation of a new, more stable cell—the eukaryote (Gr. with kernel).

Individually, archaeal and bacterial cells are constrained to be small and simple. Below a certain ratio of surface area to enclosed volume, outer membranes are incapable of supporting chemiosmosis sufficient to meet the energy demands of larger and more complex systems. Combined, the archaeal and bacterial endosymbionts contributed membrane layers with total surface area sufficient to generate the chemiosmotic energy necessary for the biosynthesis of functionally expanded complexes. The functionalities of archaea and bacteria had thus been integrated within the physically-infolded, functionally-expanded symmetry space of the eukaryotic cell; a basic attribute of all fungi, plants, and animals on earth.

Conditional reconstruction analyses indicate that the archaeal symbiont contributed primarily informational genes to the eukaryote, which were involved in transcription and translation, whereas its bacterial mate contributed operational genes associated with energy metabolism and lipid biosynthesis. This union of archaeal and bacterial endosymbionts provided the eukaryote the means to implement: 1) phagocytosis, the capability for one cell to engulf, or eat, another cell to assimilate nutrients, 2) the acquisition of genomic functionality, and 3) the sequestration of informational DNA within a nuclear membrane, separated from the mitochondrial DNA that needs to be dispersed throughout the cytoplasm external to the membrane in order for it to maximize the speed of local energy production (respiration). Archaea and bacteria initially branched on the phylogenic *Tree of Life* through conventional Darwinian selection, shown in Figure 4-3, and later reunited through endosymbiosis. The eukaryote had thus emerged on an evolutionary pathway appropriately termed the *Ring of Life;*[20] shown in Figure 4-4

The evolutionary process of endosymbiosis, by which the functionalities of existing cells merged within a new and more versatile single cell, was complemented by another method of complexification, referred to here as *exosymbiosis,* by which fitness was improved through networking of same-genotype cells—the development of multicellular complexes was most efficient when the specialization and cooperation of individual cells occurred among self-replicated, same-species cells having a consistent genetic identity. Self-replication of kindred cells spawned progeny that interacted with each other. Within a colony, local groups of self-replicated cells start out with the same DNA-based developmental potential. But, in a process called "inductive interaction," signals from cells outside a local group may drive one or more of the members inside the group onto different developmental pathways, leading to functional versatility.

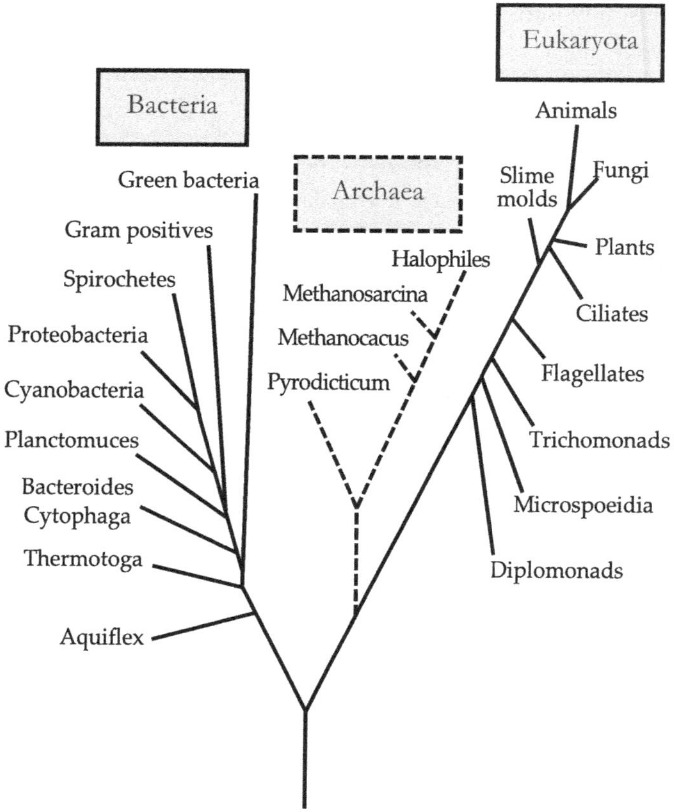

Figure 4-3 *Tree of Life*

Some inductive signals are short-range, and are transmitted via cell to cell contacts, while other inductive signals are long-range, and mediated by molecules that diffuse through the extracellular medium. Based upon this mechanism of intercellular communication, unicellular organisms established symbiotic relationships that bestowed collective properties beneficial to each member of the colony by providing them with a safer, more reliable local environment. For example, by bonding together, groups of cells presented a collectively larger size to predators, creating

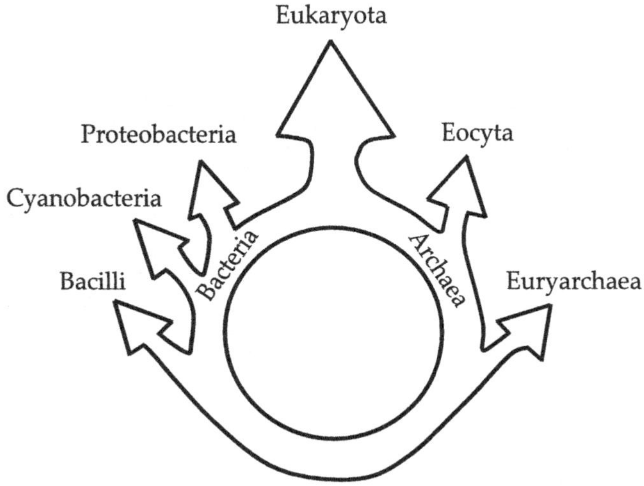

Figure 4-4 *Ring of Life*

a deterrent that protected each individual cell. Moreover, symbiosis among cells cultivated specialization for division of labor, storage of reserves, and so forth. These improvements facilitated the discovery and adaptation of increasingly complex cooperative structures and hard-wired interdependency pathways, which subserved emergent, novel functionalities of maximum benefit to the infolded colony as a whole. The essential role of cooperation in the evolution of life on earth has been succinctly captured in the words of Margulis and Sagan:

> *Life did not take over the globe by combat, but by networking.* [21]

The trend to select for infoldment continued within the holistically connected colonies of specialized cells. Spatially, functionalities were upgraded through the concentration of specialized cells into tissues and organs; and temporally, through conduits assembled for faster resource and information transport. The ability of cells to diversify synergistically depended on the exploitation of a common genotypic structure, which took the form of a DNA molecule capable of expressing any functionality encoded for the fitness of the multicellular phenotype. As the result of exosymbiosis, colonies of unicellular organisms had collectively invented the holistic integrated symmetry space of the multicellular organism. The *colonial theory of multicellularity* explains how unicellular organisms gathered into colonies, differentiated to promote symbiotic relationships, and interconnected to form holistic multicellular organisms possessing emergent capabilities and functionalities. The origin of multicellularity thus included interstitial communication, cell to cell adhesion, inductive interaction, and an alignment of fitness among cells locally, as well as a global export of fitness from individual cells to the holistic multicellular organism.[22]

The ancient, but still extant slime mold provides an interesting example of a lifeform that evolves from a symbiotic colony of individual cells into a multicellular organism, but may revert to unicellularity in response to environmental conditions. When food is abundant, cells act independently, multiplying by direct cell division. But when resources become scarce, the cells emit chemical signals that cause them to coalesce and differentiate, forming the interdependent parts of a holistic multicellular organism consisting of an upright stalk and a head containing spores. The spores are subsequently released, to be dispersed by the wind to fertile environments, where they germinate and begin the cycle anew.[23]

The differentiation and specialization of a cell type in a symbiotic colony is encoded into its DNA sequence by means of processes that transpire very rapidly. This can be seen in the laboratory experiment involving cultures of the *Rhodobacter* bacteria, organisms with the ability to metabolize either chemotropically or photothropocally.[24] When these bacteria are exposed to total darkness or total light, the entire population selects the most efficacious metabolic strategy after only a few days. Random mutations of their individual DNA molecules, due to coding errors, thermal disturbances, ultraviolet light, and other chance encounters are sufficient to enable the environmentally fittest bacteria (to light or darkness) to be quickly selected for reproduction. Early biotic earth was dominated by rapidly reproducing bacteria, which, later elevated by endosymbiosis with archaea into eukaryota followed by exosymbiosis into multicellularity, resulted in organisms able to complexify while keeping reciprocative pace with an ever-changing environment.

> *Becoming multicellular opens the gateway for all sorts of remarkable innovations that would be impossible for single cells. By being larger and by being made up of numerous cells, organisms can have a division of labor and also respond to their environment in new and sensitive ways, all adaptations that have led to their success.*
>
> — John Tyler Bonner [25]

Multicellularity made numerous advantages possible for living systems, including increased physical size, crucial in certain competitive environments, as well as increased complexity obtained through differentiation in interdependent cellular lineages. The transition to multicellular organization was an inducible response to environmental stimuli. Driven by environmental pressures and facilitated through inductive interaction, cellular organization proceeded in an upward spiral of functional complexity—"cellular diversity can evolve easily when functionally called for by selection." [26] Through successive rounds of cellular interaction and the establishment of complex interdependencies among differentiated cells in a multicellular colony, global attributes and integrated expressions arose in the form of holistic multicellular organisms.

In the phylum *Porifera* (sponges), the most primitive of all animals, and the only animal without a nervous system, functional integration relies on the cooperative actions of specialized cell types. For example, the beating of flagella by choanocyte-type cells in a sponge—a trait inherited from their pre-colonialized unicellular choanoflagellate ancestors, which facilitated locomotion for them—impels nutrient- and oxygen-carrying water into a central canal, where amoebocyte-type cells attach to the nutrients and deliver them to various other cell types located throughout the body, while contractions of muscle-like myocyte-type cells on the surface alter the diameter of porous openings to regulate the outflow of wastes and carbon dioxide. The relative numbers of functionally differentiated cells comprising these species of Porifera are optimized for overall fitness in accordance with a holistic Lagrangian factor that operates primarily on an available stock of transformable and totipotent (undifferentiated) archaeocyte cells. By contrast, the more competent glass sponge *Rhabdocalyptus* is a form of Porifera that performs operations through the transmission of electrical signals. It can arrest the beating of its flagella when touched by potentially dangerous particulate in its inflowing water stream by disseminating electrical signals to its flagellar cells via a hardwired network of thin extracellular cytoplasmic threads draped around a silica, "glass," skeleton. The intracellular and electrical signaling capabilities of Porifera marked a milestone in the ascent of life. But Porifera lacked a nervous system, the foundation upon which mental consciousness would emerge.

* * *

5

The Nervous System

Multicellular organisms can be distinguished by the presence or the lack of a nervous system. An organism without a nervous system is relegated to having a loosely integrated internal symmetry space in which processes are dispersed, duplicated, and operate in slow motion. Although the process cycles that facilitate the essential functions of non-neural organisms (such as the modulation of hormonal secretion levels to attune to long-term circadian and seasonal rhythms, the adjustment of nutrient assimilation rates to satisfy the immediate needs of biosynthesis and metabolism, etc.) are synergistically integrated to optimize holistic fitness, inter-causal signaling among these cycles is relatively slow and poorly synchronized. The entire phylogenic domains of plants and fungi function on the basis of such spatiotemporally dispersed symmetry spaces.

Originally, certain pre-neuronal cells secreted metabolites and small peptides into the inter-cellular medium. These chemicals functioned as messengers, guided by the strongest of the nonlinear electrical action potential patterns that were superimposed upon the extant network of symbiotic cell-to-cell interactions. For example, Cnidarian obelias formed colonies of polyps to constitute a shared digestive system. Electrical signals spread from one end of the polyp colony to the other, traveling along differentiated epithelial cells lining the interior of their collective digestive cavity.[27] In another example predating nervous systems, a complex protein scaffold was present in certain types of Porifera. Other nervous system precursors in Porifera included impulse conduction and osmophilic regions.[28] Although Porifera did not express neurons, they did possess genes that are today expressed by the human genome in the development of its nervous system.[29] Note that some phylogenists propose that Porifera once had a basic nervous system but later lost it, and that the branching of Porifera on the metazoan tree occurred between Ctenophora and Cnidaria,[30] not before; however, this unresolved technical issue does not abrogate the purport of our discussion as it concerns the underlying principles of neural evolution.

The specialization of the neuron cell to conduct and convey electrical signals heralded the age of the nervous system. Nervous systems evolved quickly upon the platform of nonlinear action potential patterns (which were implemented through electrochemical devices such as the *voltage-gated sodium channel*), globalizing the organism's intercellular electrical signaling network.[31] With the first appearance of nervous systems 570 million years ago, the selective efficacy of multicellular animals increased dramatically. Faster signaling and better functional coordination within the bodies of these animals were achieved through the adaptation of cells into neurons dedicated to transmitting and integrating information.

In the Introduction, the evolution of the integrative interneuronal domain was described—bifurcation of mechanosensory-contractile cells into interactive sensory neurons and interneurons, along with coordinated long-range messaging, via principal neurons, through temporal alignments among distributed interneuronal control circuits.[32,33] Over time, principal neurons elongated and branched, innervating mutually evolving non-neuronal tissue systems, thereby shortening information transmission times between interdependent distal regions of the body. The faster inter-cellular interactions that were made possible through the agency of neurons brought functional interdependencies into *temporal focus,* enabling more coordinated responses of the holistic organism to both internal and environmental stimuli. The signaling currency of inter-connecting neurons also evolved. Specialized messenger chemicals (neurotransmitters) adapted to convey informational signals through tiny clefts (synapses) between proximate neurons with speed and diversity.

The earliest nervous systems consisted of decentralized networks of principal neurons, or nerve nets, which interconnected widespread regions of the body. Primitive nerve nets extended throughout the bodies of animals in the phyla *Cnidaria* (e.g., hydra) and *Ctenophora* (e.g., comb jellies), enabling them to respond holistically to their internal states, and (through the adaptation of receptor molecules on cells located at the horizon boundary separating the microcosm from the macrocosm) to sense contact with the external environment. The nerve net improved the coordinated responsiveness of these organisms, albeit through use of the same motor response irrespective of the point of sensory contact. Though the transfer of information through nerve nets was faster than through prior non-neuronal cell-to-cell electrochemical, bulk material, hormonal, and cytoplasmic conduction transport systems, the functional coordination and overall responsiveness of these neuronal animals to their environments were limited.

Neuronal efficiencies improved through differentiation into functionally specialized cell-types such as sensory and motor, and through cellular elongation and branching for fast, distal, and parallel signaling. This first level platform of neurons established a hard-wired, global relay network that provided selective advantage to the organism as a whole. In the interior of the organism, local (constructive/destructive) electrical and (excitatory/inhibitory) chemical interferences among the first level neurons self-organized into a second-level system of signaling activity. In response to environmental pressures affecting the adaptive holistic organism, the second-level solidified through the specialization of hard-wired interneurons that *regulated* the underlying first-level network. Efficiencies gained from the second-level excitatory/inhibitory networks of interneurons were further enhanced by neuronal consolidation into local groups of ganglia and nuclei, and by the fasciculation of neurons into bundled tracts of parallel fibers which formed nerve cords to avail damage control redundancy for the functionalities of individual fibers, and to adjust experience-driven neural network pathway bandwidths and routing.

Over time, heightened neuronal efficiencies were attained through the continuing process of cephalization which, through the projection and tight interweaving of nerve tracts, facilitated rapid and collaborative interactions among closely neighbored functional plexuses. The hydra (*Cnidaria Hydrozoa*) possess a nerve net that extends throughout the body. In some species of hydra, parts of the nervous system exhibit a more advanced degree of complexity acquired through the process of cephalization. These hydra have a head and mouth region infolded within a nerve net that includes two categories of specialized neuron cells: photoreceptive sensory cells, and proto-ganglionic motor cells involved in tentacle function.

The radial nervous system of the starfish *Echinodermata Asterozoa* is somewhat more complex than that of the hydra. Starfish (also called sea stars) incorporate dedicated sensory and motor nerve nets in each of their (five or more) arms, all of which are interconnected via a central circumoral nerve ring that contains its own nerve sub-network.[34] In the more evolved species of starfish *Archaster typicus*, the ganglia concentrated in the circumoral region are able to coordinate distal sensory neurons with the feeding movements of the mouth. This starfish also demonstrates global coordination of motor function: if a sensory receptor on a particular arm detects an odor, top-down signaling from the ring is initiated, via its motor nerve net, causing its tube feet and musculature to move the mouth region toward the sensed prey. Another trait adapted by this starfish, utilizing its ring/radial neural architecture, is its global control of balance and directional systems.

The spatiotemporal infoldment of centralized, fast-signaling, neural processing centers made possible the parallel, collaborative development of elaborate sensory organs, such as eyes, and motor functions, such as mastication. The rag worm *Platynereis dumerilii* has eyes that enable it to explore its environment and react as directed by its centralized "mushroom body" brain. Genotyping analyses indicate that the same genes that express ontogenetic development of this annelid's primitive brain also express the neocortex of the human brain, indicating an early, common genetic origin, before the phylogenic branching of reptiles and mammals.[35]

Neuronal organisms are extremely attuned to their external environment (the macrocosm). Behavioral concordance relative to macrocosmic factors results from the efficiencies offered by the rapid, coordinated neuronal information processing capabilities that operate on the basis of neural Darwinism (see the Introduction for a description of neural Darwinism). With respect to the internal environment (the microcosm), prior to the postnatal phase, the organism becomes highly optimized and uniquely customized to events guiding its ontogenesis. The development of the human nervous system, for example, starts with an overproduction of neuronal stem cells. These pioneer cells project randomly into their microcosmic environment, which consists of other developing tissue types with which they interact, producing a hard-wired infrastructure reflective of their actual encounters. Overproduction and hard-wiring is followed by the elimination of the surplus neurons not activated during the environmental interface. In compliance with the principle of neural Darwinism, the fittest of neuronal populations have survived. Moreover, in the developing embryo, entire ganglionic complexes and nuclear modules are also overproduced, offering higher levels of selective opportunity for division of labor, parallel processing, backup security, and standardization of modular mechanisms.[36]

The ultimate achievement of neural Darwinism concerns the nature and functionality of the most complex physical structure known. This holistic instrument has many transcendent capabilities. It is able to apprehend and integrate information originating both in the external macrocosm and the internal environment of the microcosm to which it belongs. It is capable of enriching that information with dimensionalities of purpose and meaning relative to its own dynamic self. It engenders a first-person perspective on that information, and is endowed with a faculty that enables it to experience things, ideas, and feelings in a world of phenomena. It is the pinnacle of evolution; and the subject of the next two chapters, where we explore the marvel of the universe—the human brain.

* * *

6

The Physical Human Brain

With hundreds of functional neuronal clusters, interconnected through trillions of synapses, packed into a one quart volume, the human brain epitomizes a cephalized, spatiotemporally focused, integrative symmetry space. It is the most infolded and complex structure known. The architectonic evolution of the brain—and, in general, the nervous system and all other bodily subsystems—can best be appreciated as the progressive coupling and nesting of ever-emerging internal process cycles interacting holistically within the horizon boundary of a living microcosm.

Cycles are the principal organizational motif at all levels of the brain. Throughout the course of brain evolution, newly discovered and selected process pathways have cross-connected bidirectionally within and among many functional regions by implementing a selection mechanism that has produced a dense tangle of horizontally-coupled and vertically-nested cyclic interrelationships. Because of its cyclic architecture, the brain is not amenable to a precise description in terms of straight-line flow diagrams. But notwithstanding the inadequacy of linear cause and effect flow diagrams to represent the neurophysiological complex, numerous pathways and structures have been identified that do correlate positively and informatively with certain functional interdependencies of the brain. It should also be noted that the following description of the brain's physiology is based on current neuroscience, and that much more remains to be learned. Nevertheless, the body of scientific evidence currently available does disclose the fundamental principles of brain evolution and operation, and exposes the basic framework upon which human consciousness has emerged.

The evolution of the nervous system, from the simple nerve net of a primitive hydra to the fractal weave of cycles that constitutes the human brain, was rapid. Complexity advanced quickly upon the adaptable, plastic, informational platform of the nervous system, where, through neural Darwinism, random cross-projection of neurons among process-structures promoted the discovery and infoldment of selectively efficacious interdependencies and emergent functionalities.

Cephalization of certain neural functionalities—originally carried out in distributed regions of the body in ancestral forms of life—into a centralized location, facilitated the increased processing and integration efficiencies that uniquely distinguish the brain. As noted in Chapter 5, even with a limited degree of cephalization of neural ganglia centralized about their circumoral region, the radially symmetric nervous systems of hydras and starfish provided enhanced food sensing and eating proficiencies. Bilaterally symmetric architectures enabled cephalization hierarchies able to support higher functional efficiencies, integration pathways, and central processing hubs. For example, the primitive, but still existing, bilaterian nematode *Caenorhabditis elegans*[37] has a brain cephalized at the midpoint of its body, and the flatworms of the phylum *Platyhelminthes* possess both chemosensory and photosensory systems connected to a brain cephalized at its anterior end (relative to the forward direction of its motion, where contact with food sources and danger signals are most likely to occur). This arrangement ensures swift and coordinated responses of the bilaterian as it negotiates its environment.

Cephalization of neural complexes within the nervous systems of the more advanced bilaterian vertebrates, such as reptiles, fish, birds, and mammals, transpires in conjunction with the sequestration of the functionalities they specifically support. Selective maximization of process efficiencies for the complexes supporting the most salient functionalities and degrees of informationl integration induced these complexes, via cephalization, to rise through the spinal cord, the brain stem, and the hierarchy of the cortex.[38] For example, in fish, the highest levels of visual information processing and integration are located in the midbrain; but in mammals, these functions have cephalized to the forebrain. With the cephalization of existing functionalities, the discovery and integration of new ones becomes practicable. In particular, the integration of multimodal information presents fitness-improving traits to an adapting microcosm, both with respect to the external environment as well as to its internal physiology.

ANATOMY OF THE BRAIN

We begin this section on the anatomy of the brain by presenting a list that itemizes the major components, functions, and connections of three process-structures of foremost significance to mental consciousness: the thalamocortical system, the limbic system, and the basal ganglia; see Table 6-1. This list is provided to introduce these systems, with more detailed descriptions to follow. Note the large number of functions and connection targets shared among these systems.

Cephalization of the human nervous system has produced a hierarchy of four contiguous regions; namely, the spinal cord, the rhombencephalon (hindbrain), the mesencephalon (midbrain), and the prosencephalon (forebrain); see Figure 6-1.[39] The rhombencephalon is further partitioned into three sub-regions, the medulla oblongata, the cerebellum, and the pons, which are concerned with body movement, blood circulation, respiration, and other visceral functions. The mesencephalon has two partitions, the inferior and superior colliculi, involved in visual and auditory processing. At the top of the cephalization/infoldment hierarchy is the prosencephalon, which is further subdivided into the diencephalon and the telencephalon.

Table 6-1 List of Neural Formations

1 - Thalamocortical System

a - Cerebral Cortex [1]

Component Areas (Brodmann Areas)	Functions
1, 2, 3 - Primary somatosensory	Sensation, motor learning
4 - Primary motor	Motor movements
5 - Somatosensory association	Spatial imagery in deductive reasoning, language processing, visuomotor attention
6 – Premotor, Supplementary motor	Sensory guidance of movement, Speech perception, word retrieval,
9 - Dorsolateral/anterior prefrontal	Motor planning and organization, sustaining attention, working memory
10 - Anterior prefrontal	Memory retrieval and executive functions, reasoning, language, working memory
17 - Primary visual (occipital lobe)	Map of spatial information
22 - Primary auditory	Language and auditory processing
23, 24, 26, 29-33 – Cingulate	Emotion, learning, memory
37 - Occipitotemporal Gyrus	Face and body recognition, word and number recognition
39, 40 - Wernicke's area	Language comprehension
44, 45 - Broca's area	Motor speech programming

1 – General functions of cerebral cortex: neural integration, attention, perception, thought, memory, language, consciousness.

....

b – Thalamus [1]

Nuclear Group	Functions	Connections
Anterior (part of limbic system)	Memory, association	Hippocampus, cingulate cortex
Ventroanterior and Ventrolateral	Learning and planning of movement	Motor cortex, basal ganglia, cerebellum
Dorsolateral	Emotion, behavior	Visual, parietal cortices
Dorsomedial	Memory, attention, thinking	Prefrontal cortex, limbic system

1 – General functions: regulation of consciousness, sleep, alertness, motor and language systems.

General connections: sensory signal gateway, associated cortical areas.

2 - *Basal Ganglia* [1]

Components	Functions	Connections
Striatum [2]	See note 2	Prefrontal cortex, amygdala
Pallidum	Motor	to motor, thalamus
Substantia nigra	Motor, reward	Thalamus
Subthalamic nucleus	Action selection	Motor cortex

1 - General functions of basal ganglia: neurotransmitter value system, action selection, voluntary motor, learning, cognition, emotion;

General connections: cerebral cortex (prefrontal – executive, orbitofrontal), thalamus, brainstem, limbic circuit; gate to working memory.

2 - Includes caudate nucleus, putamen, nucleus accumbens, olfactory tubercle.

Connections: afferent from other brain areas, efferent to all other basal ganglia areas.

Functions: modulates feedback signals from cortex to thalamus.

3 - Limbic System [1]

a – Cortical Structures	**Functions**	**Connections**
Hippocampus	Memory: spatial, long-term, cognition	Entorhinal cortex
Entorhinal	Memory, association	Interface between hippocampus and other brain structures
Orbitofrontal (in prefrontal)	Decision making	Thalamus, amygdala, hypothalamus
b – Subcortical Structures		
Amygdala	Emotion, semantic memory, attention, reward system	Hypothalamus, thalamus, olfactory cortex
Hypothalamus	Autonomic processes, memory, attention	Brainstem, amygdala, hippocampus, frontal lobe

1 - General functions of limbic system: olfaction, visceral processes, emotion, motivation, long-term memory.

....

The diencephalon is divided into the hypothalamus, a collection of nuclei involved in homeostasis, sexuality, and emotion; and the thalamus, a relay station to higher sensory functions; see Figure 6-2.

The telencephalon is the embryonic structure from which the basal ganglia and the cerebral cortex develop. The basal ganglia and cerebral cortex, along with the hippocampus and olfactory bulb, comprise the phylogenically newest complex of the brain, collectively referred to as the cerebrum.

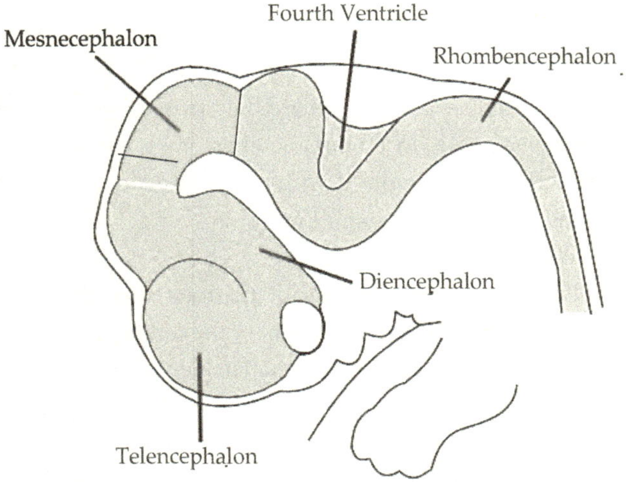

Figure 6-1 *Cephalization of the Human Embryo*

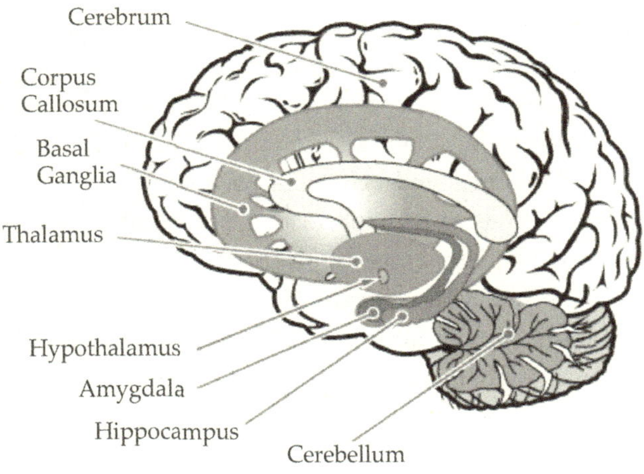

Figure 6-2 *Subcortical Process-Structures*

The basal ganglia are a collection of interconnected nuclei mostly involved with telencephalic functionalities, but they are also crisscrossed with more primitive encephalic processes. They are common to all vertebrate brains, from the primitive lamprey where they primarily accommodate the basal motor behaviors, to mammals where they contribute to the executive-motor repertoire involved with action selection,[40] reward and learning systems, emotion, and cognition.

The cerebral cortex consists of layers of neurons specialized for the integration of information. The latest region of the cerebral cortex to have evolved is the highly partitioned mammalian neocortex, with human beings having the most complex of all. Indeed, there are 500 billion neurons in the human brain, interconnected by an average of seven thousand synapses on each, all merged together into a self-modulating, multilevel neural network. In the cerebral cortex alone, there are 100 billion neurons,[41] one quarter of which are interneurons specialized for integration.

The neocortex operates with support from a host of subcortical process-structures which it covers dorsally. Collectively, the basal ganglia, thalamus, hippocampus, and numerous other subcortical structures help to regulate the integration of information in the neocortex which governs advanced functionalities and behaviors, including abstract reasoning and social interaction. In addition to the subcortical structures listed, mammals possess another group of functional structures, including the amygdala in the limbic system; see Figure 6-3. The limbic system is intricately enmeshed with other cortical and subcortical cycles, and is essential to functions of memory, emotion, and motivation.

The development of the human nervous system from its embryonic origins begins with the differentiation of neuroepithelial cells, which form a region of tissue called the neural plate. The plate folds and extends into the neural tube, which fills with cerebrospinal fluid. A mature neural tube develops a structure of connected compartments, consisting of the spinal canal below, the medulla oblongata, and a series of four chambers above, called the cerebral ventricles. While this basic linear arrangement of components along the neural tube has remained conserved through evolution, selective influences have come to build other structures that embellish complex brains. Randomly projected neurons, from and into functional complexes, which intimate improved holistic fitness, invoke pathway entrenchment. Through this method of discovery, emergent systems cephalize and infold for efficient interactivity and capability, allowing functionally integrated symmetry spaces to arise.

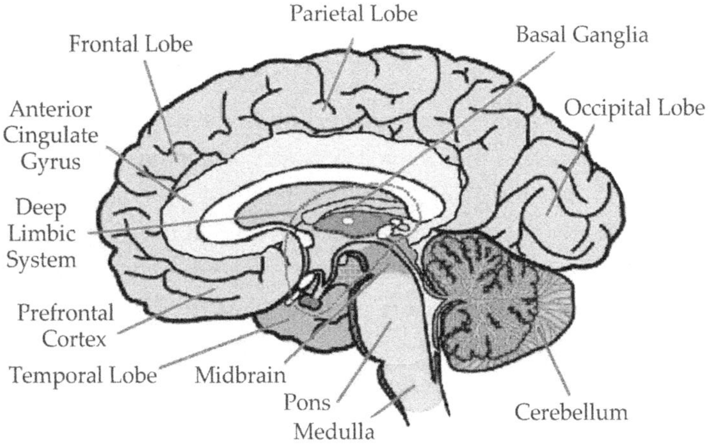

Figure 6-3 *Cortical Process-Structures*

Chapter 6

TRIUNE BRAIN

In Western philosophy and science, consciousness has long been equated with the higher mental faculties of human thought, feeling, and volition. Creatures with brains simpler than those belonging to human beings, such as reptiles and birds, have been considered by many not to be conscious. In the 1960s, this opinion was quantified by the *triune brain theory*,[42] by setting the threshold for consciousness to be the possession of a neocortex, the exclusive instrument of higher mental cognition and abstraction; for instance, numerical mental computation and quantitative conceptualization jointly involve the temporal, parietal, and prefrontal regions of the neocortex.[43]

At the time when the triune brain theory was formulated, it was believed that three regions of the brain had evolved chronologically. In Chapter 5, however, we noted that the same genes which express biosynthesis of neocortex in the human brain also express the mushroom body brains of primitive annelids, indicating a very early common origin, before the phylogenic branching of reptiles and mammals. However, because information processing competency is proportional to the number of layers comprising cortical tissue, and layering has generally increased over time, along with the complexity it enables, the triune evolutionary tree does roughly parallel phylogenetic evolution. Triune brain theory is informative and historically interesting in this important context: the genetic expression of brain fitness actualized by selection. Therefore, a brief review of triune brain theory is presented, followed by an update to incorporate recently discovered genetic implications.

According to the triune brain theory, three distinct complexes derived from the prosencephalon through the course of vertebrae evolution. The earliest was the reptilian complex (500 million years ago), associated with allocortex (archicortex and paleocortex, tissue of 3 to 5 layers) and subcortical basal ganglia. The paleomammalian complex was the next to self-organize (150 million years ago), in the form of the limbic system. The limbic system is comprised of the olfactory cortex, septum, amygdala, hypothalamus, and hippocampus. The most recent system to evolve (3 million years ago) was the neomammalian complex, which includes the six-layer neocortex. The relative amount and anatomical distribution of the cortical tissue types, as they appeared through the course of evolution, is illustrated in Figure 6-4.[44]

The archicortex is a three-layer structure common to mammals and pre-mammalian vertebrae, suggesting the premise of the triune brain theory: that the archicortex, supported by the coevolving subcortical septum (shown in the figure) was an evolutionary predecessor to the neocortex. The archicortex is lined by a variety of functional structures, including the amygdala, which modulates memory consolidation of what has now become emotion; and the hippocampus, which both coordinates spatial navigation and consolidates information sent from short-term to long-term memory.

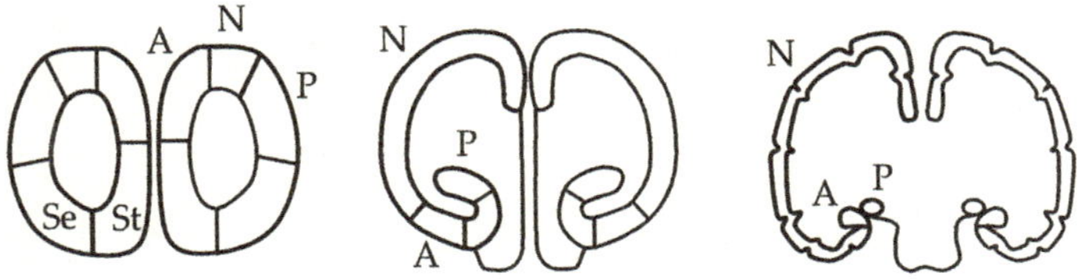

Evolution of cortex through a lower vertebra (shark), a mammal (rat), and a human

A - Archicortex
P - Paleocortex
N - Neocortex
Se – Septum
St - Striatum

Figure 6-4 *Cortical Process-Structures*

The hippocampus consists of a profusely-branched dendritic arbor of synaptic interconnections capable of encoding spatial representations of a microcosm's external environment, a trait of paramount importance to the survival of most advanced animal species. Interestingly, the three-dimensional coordinates delineating the spatial configuration of the external symmetry space are not mapped directly onto a corresponding three-dimensional array in the hippocampus; rather, a geographically reorganized synaptic strength density pattern encodes the configuration of the external world.

The paleocortex is primarily concerned with olfactory function and, as portrayed in the figure, has diminished in terms of relative salience. Most recently, mammals appeared with their six-layered neocortex. The neocortex evolved in conjunction with a contingent of supporting subcortical structures and process systems, including the thalamo-basal ganglia-cortical system which now serves the functions of higher, focused consciousness. In the paleo-cortical era, the simple structure of the hippocampus provided the brain its primary cerebral memory space. In the human brain, certain functions of the hippocampus have been subsumed by the neocortex, in the sense that earlier input and output connections of the hippocampus have been transferred to the neocortex. As such, the memory function of the hippocampus is now an appendage to the thalamocortical system, which connects directly to the external environment.

The triune theory depicts brain evolution as the sequential advance of cortical complexity with respect to the expressed phenotypes of animals. But, the underlying genetic and developmental processes responsible for the expression of these various evolutionary stages have been precisely conserved through brain evolution. As noted above, the genetic heritage of the mammalian neocortex dates from pre-reptilian annelid mushroom body brains. In the development of both reptilian and mammalian brains, a cortical pre-plate forms in the telencephalon which implements the general process for producing cerebral cortices. It has been shown that cortical adaptations are concerned with *operational* proficiencies established through standardized neuro-architectonic devices, such as modular componentry, layering, and integrative expandability.[45]

FRACTALS

The discovery that cortical componentry contributes operational capacities, rather than complete and specific functionalities, is consistent with the observation that brain evolution has ascended a fractal hierarchy (similar at all scales) that encompasses both geometric structures and cyclic process-structures. The fractal modeling motif underlying neural architectonics represents a fundamental and pervasive feature of evolution that, in the abiotic domain, governs the successive bonding of elementary particles, atoms, and molecules, as well as the astrophysical formation of planets, stars, and galaxy clusters.[46] In holistically closed biotic regimes, fractal composition permeates the pathway networks of biocatalytic sets, the tissue structures and macromolecular folding hierarchies of organic cells, and the coupled and nested neuronal cycles of the brain.

The fractal evolution of the nervous system has unfolded on the platform of electrochemical interactions that bond the atoms of the periodic table into structures with incongruous shapes and disproportionate electronegativities. The consequence of these incommensurable conditions is that interaction processes involving atoms, molecules, and neurons are nonlinear. As described in Chapter 3, open-system biocatalytic sets were able to self-organize upon the foundation of nonlinear electrochemical properties.[47] The primitive, holistically-closed biocatalytic set has since evolved, fractally, into the complex nonlinear symmetry space of the human brain.

The evolution of fractal symmetry spaces has undergone a series of abrupt phase transitions (changes of dimensions and parameters) corresponding to emergent levels of organizational closure. Phase transitions in the microcosm have proceeded through layered states of *self-organized criticality,*[48] toward an attractor of holistic complexity; a symmetry space with a Lagrangian that exhibits spatiotemporal scale-invariance and self-similarity with respect to enfolded levels of cyclic process-structure, including the causally affective neural levels of the brain.[49]

> *During the course of evolution, the basic circuits of the brain are not fundamentally modified, but instead, multiple parallel circuits, consisting of intermediate and longer chains of neurons, are superimposed on the existing wiring. No matter what fraction of the brain web we are investigating, neuronal loops are the principal organization at nearly all levels. This multilevel, self-similar organization is a fractal of loops.*
>
> — György Buzsáki [50]

Buzsáki identifies the organization of the brain's process-structures as a multilevel, self-similar fractal of loops. In the cerebral cortex, process-structures consist of a hierarchy of convergence-divergence complexes linked by bidirectional loops (cycles). Rising through the organizational hierarchy of the cortex, we find: pyramidal neurons with intra-cycling branching structures; cortical columns in which thousands of pyramidal neurons are clustered, projecting feedback and feedforward trajectories; recurrently linked columnar complexes comprised of cortical columns; and, at the nodal level, reciprocative patches of columnar complexes. At the highest, transmodal level of the fractally organized nodal network are the cycles which may instantiate integrated informational representations and dual collective structures in the dynamic core of mental consciousness.

APPENDAGE AND VALUE SYSTEMS

Because of their specialized functionalities, the cortico–basal ganglia–thalamic system is often partitioned into the thalamocortical system and the *cortical appendage system*. The cortical appendage system, referred to above as subcortical process-structures (basal ganglia, cerebellum, hippocampal complex, etc.), consists of bidirectional loops that cycle to the cortex (sometimes via the intermediating thalamus) and back. The components of both the thalamocortical and cortical appendage systems are innervated by global processes that comprise the *ascending value system*. The ascending value system consists of a diffuse spray of projections from the brainstem and the hypothalamus, which innervate the entire brain. The value system releases neuromodulatory chemicals, including serotonin and dopamine, which signal types of salience, such as a loud noise, a sudden pain, or a mood of pleasure; see Figure 6-5. The fan-like projections of this neurotransmitter-based value system also participate in the plastic reorganization of cortical and subcortical synaptic memory structures.

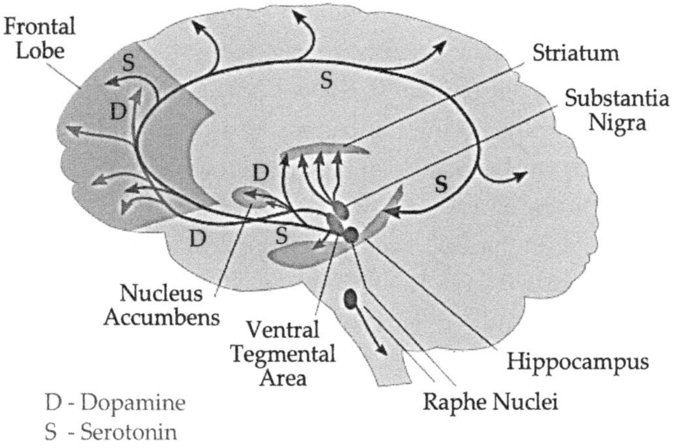

Figure 6-5 *Neurotransmitter Value System*

Chapter 6

THALAMOCORTICAL SYSTEM

Owing to its direct contact with every sensory pathway connecting the cerebral cortex with the environment, its numerous connections to internal functional centers, and its resonance at 40 Hz during all events of mental consciousness, the thalamocortical system provides a logical starting point for associating the regions of the brain. The thalamocortical system consists of the thalamus (a cluster of about 50 nuclei located at the center of the brain) and the cerebral cortex (a massive swath of cortical neurons covering the thalamus from above). The cerebral cortex (neocortex and smaller allocortex, differing in number of neuronal layers) is partitioned into regions dedicated to the integration or differentiation of informational modalities. In particular, three principal regions of the neocortex, the posterior, limbic, and anterior, associate to facilitate the most advanced capabilities of the human brain. The posterior region, consisting of the occipital, temporal, and parietal lobes of the neocortical swath, integrates unimodal sensory (visual, auditory, etc.) signals into multimodal patterns crucial to human perception and language. The limbic region, comprising the anterior-ventral portion of the temporal lobe, is involved with emotional content, memory, and learning. The anterior region, primarily the frontal lobe, is concerned with executive-motor functions, working/short-term memory, high-order cognition, regulation of emotions, and the motivational reward system. The posterior, limbic, and anterior cortical association regions, along with others, are highly interconnected, and establish a unified symmetry space for the representation of integrated information.

The thalamocortical system may be visualized as a tetrahedron (a closed surface of four triangles) standing on one of its vertices. The tetrahedron's six edges represent *bidirectional* neuronal pathway signaling cycles, which may be either constructive or inhibitory, together forming a reentry network. The horizontal triangle at the top of the tetrahedron is the neocortex, which has no direct contact with the external world. Its vertices are its functional association regions. The three descending edges of the tetrahedron congregate, but do not interconnect, in the thalamus below. The edges of the neocortical triangle (corticocortical cycles) integrate its functional regions, while the edges between the neocortex and the thalamus (thalamo-cortical cycles) connect the regions of the neocortex to functionally corresponding nuclei in the thalamus. In addition to cycling with sensory association cortices, the dorsomedial nucleus also cycles with executive-motor association regions. Through the course of brain evolution, top-down executive functions, operating from within the corticocortical complex upon bottom-up sensory and emotional signals, has taken a large measure of control over the selection and modulation of pathways to the environment; there are many more neuronal projections from the neocortex to the thalamus than vice versa.

Components of the thalamocortical system also share bidirectional signaling cycles with the subcortical appendage systems. Appendage systems extend and enhance the thalamocortical system in a number of interrelated ways. For example, the hippocampus is an appendage system involved with memory management and storage; while the complex of appendage systems called the basal ganglia participates in invoking motor functions, emotional arousal, cognitive processing, and the executive control of behavior. Appendage systems participate in a triangular relationship with the neocortex and thalamus, connecting bidirectionally to each. In particular, components of the basal ganglia are connected to the thalamus, prefrontal and motor cortices, and limbic structures including the amygdala and hippocampus. Additionally, appendage systems cycle signals among themselves—just as catalytic sets in unicellular organisms tweaked and embellished the evolution of their own intra-complex pathway networks by adapting feedback and feedforward cycles in the selective search for fitness, so have evolving neuronal systems incessantly intra-projected neurons to discover and establish signaling relationships beneficial to the holistic microcosm.

The thalamocortical system is made up of a group of bi-directional processes that cycle between the thalamic nuclei and functionally corresponding cortical regions (the thalamocortical arcs), as well as among the cortical regions themselves (the corticocortical arcs). A few of the approximately 50 pairs of nuclei and sub-nuclei (mirrored in each half of the brain) of the thalamus are illustrated in Figure 6-6.[51] The anterior thalamic nuclei (AT) are involved with memory, the lateral geniculate nuclei (LG) with vision, the medial geniculate nuclei (not shown) with audition, the pulvinar nuclei (P) with motor function, and the ventral nuclei (V) with premotor function. The dorsomedial nuclei (DM) cycle throughout the brain, connecting with the olfactory cortex, the limbic system, sensory-association areas, and the executive-motor system. All thalamocortical process cycles operate in conjunction with cycling to the basal ganglia.

The interaction of the dorsomedial nuclei with the prefrontal cortex plays an important role in attention, abstraction, and working memory. Thus, in addition to being the so-called gateway to the cortex, mediating all signaling between the external environment and the corticocortical cycles, the thalamus is also instrumental in modulating the integration of information at the highest multimodal levels of the cortex.

The thalamic reticular nucleus (TR) is unique in its functionality, acting as an intermediary that filters, tunes, and modulates all of the signaling between the other thalamic nuclei and the cortex. Due to its construction—a thin complex of neurons partially surrounding the thalamic nuclei—as well as its multimodal functionality, the TR has been called the seventh layer of the neocortex, though it is the most ancient of the thalamic nuclei; see Figure 6-7.[52] The multimodal functionality of the TR has a significant effect on the process of information integration and therefore on the formation of an object of consciousness.[53]

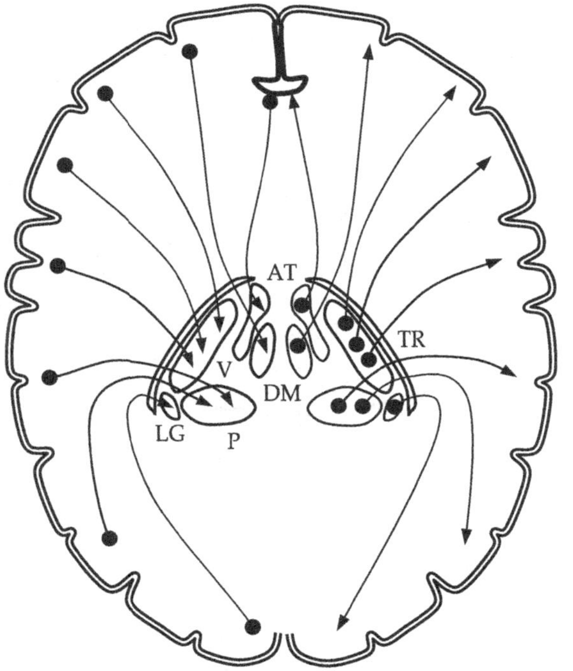

AT – Anterior Thalamic P – Pulvenate
DM – Dorsal Medial TR – Thalamic Reticular
LG – Lateral Geniculate V - Ventral

Connections between Thalamic Nuclei and Functionally Corresponding Cortical Areas.

Figure 6-6

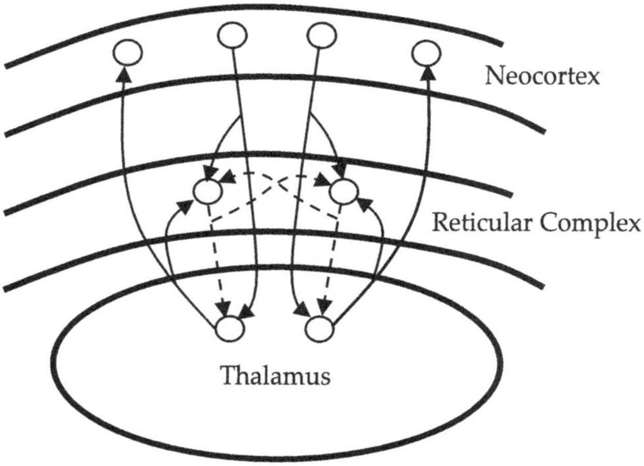

Excitatory axons are represented by solid lines, inhibitory axons by dashed lines, and synapses by arrow heads.

Figure 6-7 *Thalamic Reticular Complex*

During discrete conscious events, the gamma activity in the cortex, the thalamus, and the interconnecting circuitry of the TR is synchronized. As depicted in the figure by dashed lines, the cross-modal neurons of the TR are inhibitory, providing a gate-keeping mechanism for internal selective attention, as well as for external interactions.

NEUROTRANSMITTERS

The evolution of the functional process-structures and circuitries that support the global systems of the brain, including the thalamocortical and appendage systems, was accompanied by the complementary evolution of the neurotransmitter molecules that convey signals among neurons, as described in the Introduction. Neurotransmitters are released at the output end of a neuron (axon), travel across a tiny cleft (synapse), and are received at the input end (dendrite) of an adjacent neuron.

In the march to fitness through complexification, neurotransmitters, in conjunction with coevolved inter-synaptic transference processes, have enabled highly efficient local information communication proficiencies. Neurotransmitters also modulate global functions by inducing effects of cortical and subcortical regions of the brain via neuronal foliations that project diffusely from value nuclei, e.g., from the nucleus accumbens, ventral tegmental area, and raphe nuclei; as illustrated in Figure 6-5. Global messaging imposes salience levels and behavioral criteria upon the large-scale functioning of the nervous system. For example, the neurotransmitter dopamine, whose widespread release is triggered by actions of the ventral tegmental area, modulates motivation, whereas activation of the serotonergic projection neurons, initiated by the caudal value nuclei, participates in the regulation of emotions.

Global regulation of the nervous system by means of neurotransmitter value-leveling is a mechanism that reiterates the hormonal secretion systems responsible for regulating homeostasis and reproduction in more primitive organisms (all multicellular organisms, including plants, produce hormones). Effective performance of basic survival systems is a conserved holistic attractor that, through the course of evolution, has continually entrained emerging functional cycles into self-consistency through global network modulation. In the human brain, the most integrated neuronal system, large-scale activation of dopaminergic projections reinforces overall positive states (specifically, by enhancing dopaminergic signaling in the cortico–basal ganglia–thalamic loops of the reward system, referring to its effect on pleasure, desire, and motivation).

Neurotransmitter-based value systems mediate process cycles at all levels of the nervous system hierarchy, from plexuses, to the spinal cord, to the top of the cephalized order where the neocortex resides. The neocortex is found only in mammals. It is much more complex and functional than both the evolutionarily earlier nido/ento pallium of birds, and the three- or four-layered allocortex characteristic of the brains of fish, amphibians, and reptiles. However, all of these architecturally disparate associational structures are genetically homologous, intervene between sensory inputs and motor outputs, and function as executive crossroads to enforce essential top-down behaviors such as intention and attention. But only the most recently evolved neocortex can support the capacity to cognize abstract information and engage higher emotions like love and anger; transcending visceral homeostatic emotions like pain and hunger, and instinctual, stereotyped, or stimulus-bound brain functions, to which the neurally primitive animals are consigned.

NERVOUS SYSTEM HIERARCHY

The sensory and executive-motor pathways in the neocortex each include a hierarchy of associational regions that process highly integrated and abstracted information, very far removed from the routine sensory, skeletomotor, and autonomic functions relegated by cephalization to the lower dominions of the nervous system. In the sensory association pathways of the neocortex, bottom-up signs (electrical signal patterns) are sequentially integrated as they pass through the primary sensory, unimodal, and polymodal association regions, producing increasingly integrated informational patterns. Conversely, through the reverse top-down executive-motor hierarchy, integrated signs emanating from the prefrontal cortex are sequentially differentiated as they descend through the premotor and primary motor cortices, driving appropriate output tracts to bodily functions and the environment. Additionally, the two pathways are interconnected and inter-modulating; see Figure 6-8.[54]

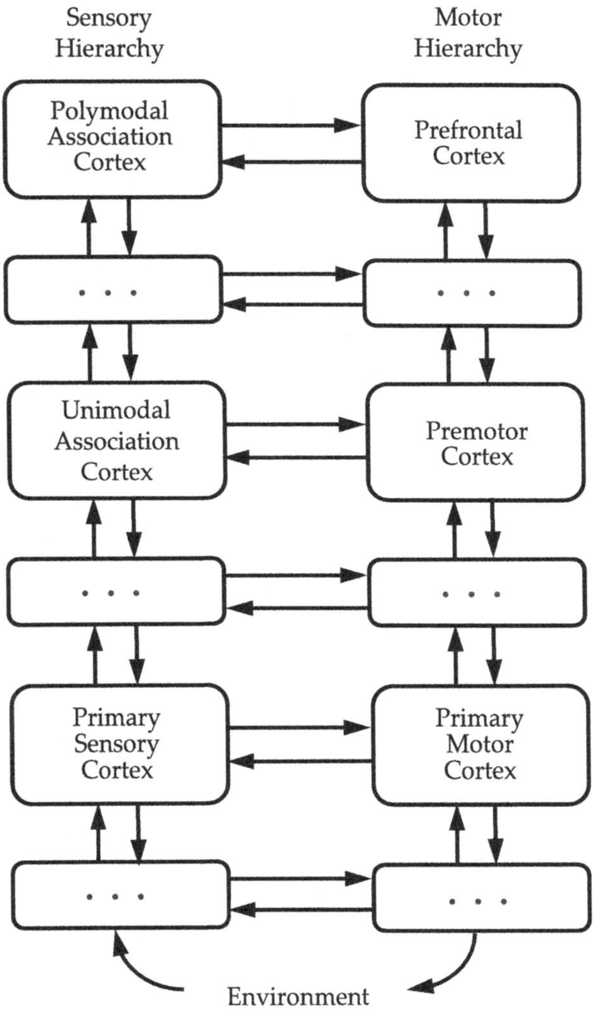

Figure 6-8 *Neuronal Fiber Connections in the Afferent-Efferent Cycle*

The prefrontal cortex resides at the top of the cephalization hierarchy, and demarcates the superordinate nexus of the nervous system's profusely crisscrossed afferent and efferent channels. Through the processes of the prefrontal cortex, maximally integrated and multimodally abstracted informational representations are apprehended in the form of temporally synchronized cycles distributed across the thalamocortical system. In turn, these resonant cycles produce and incorporate the reciprocative top-down cycles that govern the functions of intention and attention. Apprehension and intention are principal functions of the physical human brain, and are closely associated with the subjective faculty of mental consciousness, which brings us to the fascinating subject of the next chapter, the conscious human brain.

* * *

7

The Conscious Human Brain

INFORMATIONAL OBJECTS OF CONSCIOUSNESS

Experiments in cognitive psychology, aided by powerful neuroimaging technologies, have demonstrated that certain standing patterns of electrochemical activity generated in the neocortex are correlates of consciousness. Each of these patterns signifies an informational representation in the symmetry space of phenomenal percepts; viz., an object of consciousness (this chapter builds upon Sections 6 & 7 of the Introduction, where the principles underlying mental consciousness and subjective experientiality were described, and where definitions of terms such as "phenomena" and "object of consciousness" were provided). Objects of consciousness have both informational quantity and quality. The quantity equals the amount of information contained in an object of consciousness (Shannon information), in the sense of the amount of mathematical reduction in the uncertainty concerning the state of the symmetry space.[55] Quality refers to the distinguishing characteristics, or *qualia*, of the information, specifying the value actually taken (e.g., red vs. blue) in a dimensionality (color) of the phenomenal symmetry space. Sensory qualia are the most basic ingredients of an object of (mental) consciousness. The term qualia is also used more comprehensively to refer to any experienced informational representation (semantic information) that has meaning relative to a particular interpretant structure.

Qualia, and the meaningful informational representations they underlie, may thus be sensory (the redness of red, the taste of a strawberry), visceral (pain, emotion), self-referential (I versus it), intentional (the resolve to pay attention), cognitive (the notion of war), etc. In the case of the abstract word, "war," the interpreted and experienced meaning depends upon the group of neuronal cycles—the contextualized operational module—currently active in the brain. An operational module connects a *symbol* that is active in short-term memory—in this case "war"—with the meaningful nuances of war that are embedded and distributed throughout long-term memory.

MEMORY

Memory is an essential component of a microcosm's fitness, enabling it to recognize and contextualize the flood of signals produced by internal sources or coming from the environment. As with other traits of fitness, memory platforms and proficiencies have evolved throughout the ascent from unicellular to multicellular life. In a unicellular organism, memories are stored in the genetic blueprint that it utilizes for immediate exigencies signaled by its coupled phenome, as well as for passing along its remembered evolutionary heritage through many self-replicating generations. In non-neuronal multicellular organisms, the fitness trait of memory originating in each of its component cells, has been exapted to support the infrastructure that unifies those cells into a holistic network. In organisms possessing a nervous system, memory repositories based on synaptic imprinting are embedded in many types of neural complexes, as in the pallium of a rag worm, or the brainstem and cerebral cortex of a vertebra.

A diverse variety of memory systems operate and cooperate within the human body, including genetic, non-neural, brainstem, and allocortical, as well as a group of memory platforms located within the neocortex. With respect to cortical memory systems, five interrelated types are especially important to the functioning of mental consciousness: long-term memory, sensory memory, short-term memory, working memory, and the hippocampal memory system. These will now be described, followed by an outline of how they fit together in the nodal hierarchy of multimodal integrability.

In addition to their role in the global integration of salient information across the corticocortical complex, cortical neurons interconnect with each other locally to store *long-term memories*. In the same way that a vinyl phonograph record can play back musical information etched into its surface by the impression of sound vibrations, the cortex is able to encode and reproduce instantiated patterns of integrated information. Long-term memories are etched into the neural network through the strengthening, or long-term potentiation, of dendritic arbor synaptic interconnections previously activated in the construction of informational representations.

Once recorded, repeated utilization of memory pathways scribes them ever-deeper into the synaptic arrays of long-term memory. The synaptic arrays are plastic, with continuously remodeled morphisms encoding an adaptable cortical library of memories. Note that when a morphism is reactivated and incorporated into a new informational representation, that representation may or may not be allocated to the phenomenal symmetry space that subserves conscious awareness. Whether conscious or subliminal, the informational signs retrieved from memory stores have semantic meaning *only* in relation to the configuration of the particular interpretant structure in which those memories are embedded. For example, an informational sign stream relaying the image of a cat to an interpretant structure that embodies memories of a cat might evoke a feeling of fondness in a human being, but one of fearfulness in a mouse.

Another structure of importance to memory systems is the hippocampus. The hippocampus was described in Chapter 6, where it was noted that it was a precursor to the paleocortical and neocortical memory system of more recently evolved animals. The hippocampus is comprised of a three-layered archicortex. In the human brain, the hippocampus is a component of the *medial temporal lobe memory* system, which includes the entorhinal, perirhinal, and parahippocampal cortices.[56] The medial temporal lobe memory system interacts with distributed unimodal and multimodal regions of the neocortex (the frontal, temporal, and parietal lobes) to form, temporarily store, and then bind together whole, long-term, consciously recallable *explicit memories.*[57] (Explicit memory, also called declarative memory, refers to memories that are accessible to consciousness. Explicit memories are encoded by the hippocampus, entorhinal cortex, and perirhinal cortex, but are consolidated and stored elsewhere.) The medial temporal lobe memory system thus connects with regions involving sensory modalities, cognition, and high-level functions associated with the prefrontal cortex.[58] The interactions and intricate relationships of the hippocampal complex are also implicated in the mapping of spatial content and the organization of ensembles of "time neurons" which sequence the order of remembered episodes. In contrast, there are also non-consciously recallable long-term memories called *implicit memories,* as well as non-recallable *procedural memories* which support automated behaviors such as playing a musical instrument. Implicit and procedural memories do not involve the hippocampus, but are encoded and stored through operations of the basal ganglia, cerebellum, and motor cortex.

Sensory memory refers to the transient, bottom-up, stimulus-driven physical signal patterns delivered through the thalamus to the primary (visual, auditory, tactile, etc.) sensory cortices. Sensory memory is basic, resulting directly from stimuli. Sensory memory is encoded in the iconic traces of the neurotransmitters depleted in a neuronal firing. Local neurotransmitter recycling sustains this informational representation for a short period of time before it is refreshed in a neurotransmitter uptake phase, but long enough for it to be consolidated and cycled into the integration hierarchy.

Short-term memory is a more recently evolved neocortical structure that operates in conjunction with sensory memory so as to incorporate non-sensory informational patterns. Informational patterns are generally uploaded through the parietal cortex for integration, then passed on, under the control of the working memory process, to the prefrontal cortex located at the top of the cephalization hierarchy, where short-term memory representations are instantiated. Short-term memory holds representations for durations between those of brief sensory and prolonged long-term memory temporal capacities, allowing enough time for sequential bottom-up/top-down signal processing systems to interface effectively.

The data capacity of short-term memory is limited to about seven items. For example, most people can recite a number of approximately seven digits immediately after hearing or seeing it. It has been demonstrated that the storage of a number in short term memory is further apportioned into *chunks* of two, three, or four digits each. The notion of a chunk is closely related to that of a symbol, described above as a link for memory cycles participating in the instantiation of an object of consciousness. Chunking is an important cognitive trait. Studies indicate that a child's capacity for chunking correlates more directly to their future academic achievement than does IQ score.

A simple thought experiment provides an example of chunking: when one memorizes a sequence of digits, such as (π =) 3.1415926535, and then mentally recites that number, a cadenced series of 2, 3, or 4 digit chunks enters and flows through awareness, and thus short-term memory. Tasks such as reading a long sentence require maintenance of more than two or three chunks of information in an experienced representation, but short-term memory capacity would be filled after only a few words. This limitation is overcome by encoding most of what is read in distributed informational cycles, and linking them together through retrieval mechanisms based on chunked symbols. Only a few concept symbols need to be held in short-term memory concurrently, these serving as pointers by which retrieval processes link distributed cycles associated with them. The linked informational cycles interact with a co-transient collective structure to conjointly form an informational representation experienceable by the collective structure itself, on the basis of its endogenous, first-person, ultrasensory perspective.

Working memory is a processing complex, not a repository, located in the dorsolateral prefrontal cortex. It manages the contents of the short-term memory register, which is located in the adjacent ventrolateral prefrontal cortex, and is associated with long-term memory retrieval via the medial prefrontal cortex. Working memory is also an important component of executive function. It is involved in the selection and screening of the bottom-up signals, based on current contextual conditions, which may contribute to the formation of higher representations of integrated information. Reciprocally, working memory is a top-down, internally-generated reflective and modulating process that invokes goal-directed behaviors based on the contents of the integrated information.

NODAL NETWORK

The creation, storage, and retrieval of cortical memories depend on a complex network of processing nodes.[59] Physically, nodes are modular assemblies of standardized neuronal micro-columns. Nodes range from one to several centimeters in diameter. Each node resides at a specific level in a hierarchy delineated by the number of synaptic steps separating it from the point of sensory contact with the external world. Each node receives converging afferent signals, extracts and associates the pattern features, and sends diverging efferents to nodes located at the same, higher, and lower levels of the hierarchy. For example, for the sense vision, signals initiated by the impression of photons upon retinal photoreceptors in the eye are transmitted via the optic nerve to the lateral geniculate nucleus of the thalamus, which relays the signals it has processed to the primary sensory(-motor) (retinotopic) level in the cortical hierarchy for neuronal mapping of informational patterns. In turn, these first level patterns are forwarded to the second level, the unimodal association level, for the extraction of specific color, shape, and motional attributes, as well as for the encoding of integrative associations among them.

As information ascends the nodal hierarchy, complexity progressively increases through the elaboration of neuronal feedback and feedforward cycles, both within and among levels. In levels three and four, categorical information, such as seen objects and faces, or heard word-forms, is constructed. In levels five and six, different modes of sensory information, along with non-sensory information produced in paralimbic and limbic areas, are integrated. Participation of the primitive, three-layered archicortical limbic system in the construction and acquisition of high-level multimodal information (integrated patterns that unify multiple modalities) is crucial for the competitive advantage of organisms that also possess the phylogenically newer six-layered neocortex, as the limbic system enables events carrying high emotional and motivational salience to be incorporated.

Nodes located in levels five and six are transmodal (transmitting multiple modalities). Transmodal nodes are interconnected by means of self-encoded pointer directories that tie together coherent memories, thoughts, and experiences. Three important examples of transmodal nodes relevant to consciousness include: 1) the hippocampal-entorhinal system, concerned with the recording and retrieval of explicit (conscious) memories (distributed information can also be stored and recalled subliminally through the interactions of the transmodal network, but the hippocampal-entorhinal system and its closely related limbic structures are necessary for binding together unitary objects of consciousness); 2) the Wernicke and Broca areas, which support the comprehension and expression of semantic language; and 3) the short-term/working memory complex, located in the prefrontal cortex, which is central to the administration of executive functions and to the assembly and temporal dilation of the most complex informational patterns, including, but not limited to, objects of consciousness.

Nodes in the prefrontal cortex, especially the working memory nodes, play a major role in fixing attentional focus and determining response selection by promoting the flow of novel or salient informational signals up through the hierarchy, while deterring the ascension of routine, non-novel, non-salient, or predictable bottom-up information. In this respect, the prefrontal cortex is the apex and epicenter of the human brain. Apex refers to the location of the prefrontal cortex at the top of the cephalized nodal hierarchy. Epicenter refers to the consolidated nexus of the pre-frontal cortex where maximally integrated, bottom-up signs converge, and from which top-down executive signs diverge.

Interestingly, it has been established that although certain bottom-up integration pathways may be inhibited from rising to the level of the phenomenal symmetry space by top-down inhibition pathways, sub-conscious cognition may still attain a threshold of neuronal excitation sufficient to be recorded in the corticocortical meaning repository, as evidenced by the attentional blindness and other experiments. Thus, subliminal[60] integrated information is routinely produced in the non-phenomenal dimensionalities. In contrast, the information integrated *by* the group of cycles looping through the short-term/working memory complex in the prefrontal neocortex (at the pinnacle of the nodal hierarchy), correlates with the phenomenal symmetry space of mental consciousness.

The effectiveness with which informational representations are selected and integrated into objects of consciousness is the product of several evolutionary adaptations to the neural network. The offloading of functional processes onto subliminal subsystems (the peripheral and autonomic nervous systems, cognitive processors, memory storage and retrieval systems, etc.) has unencumbered goal-specific selection of the most effective pathways with which to construct objects of consciousness.

That is, the process mechanisms and resource allocations deemed optimal by natural and neural selection for production of objects of consciousness have evolved so as to ensure holistic balance between the advantage of adroit responsiveness to the demands of internal and external environments, and the benefits of fast-paced, high-content, multimodal objects of consciousness.

The description of the process of consciousness in terms of bottom-up and top-down pathways must be broadened to include their profusely interconnecting network of coupled and nested process cycles, the basis of the brain's complexity and holism. The straight-line, bidirectional notion of bottom-up/top-down pathways, running from sensory receptors to the prefrontal cortex and back, must therefore be expanded to account for informational sources located at intermediate positions in the cyclic nodal network. In our preceding description of the nodal network, the sensory first level was taken to be the starting point. However, endogenous sources, independent of sensory processing and integration, provide additional starting points. Endogenous activities originating in the frontal lobe, the limbic system, or anywhere else in the network, may be viewed as sources of both bottom-up informational flows, as well as top-down self-proficiencies including attention,[61] memory recall, and prediction of forthcoming sensory events, with novel or unexpected inputs gaining attention. In fact, top–down influences may be activated in the complete absence of any bottom-up stimuli, as shown, for instance, in mental imagery studies.[62]

In the physiological dimensionalities, bottom-up sensory-limbic and top-down executive-motor pathways inter-cycle to mediate physical processes; whereas, with respect to the phenomenal dimensionalities, bottom-up apprehension and top-down intentionality inter-cycle. For example, phenomenologically, this explains how the intention to think in words, or to communicate through speaking, results in the assembly of intelligible sentences: In response to the apprehension of integrated information in Broca's area (specialized for language production), top-down signs are broadcast as nascent pulses of intentionality. These signs correlate vocabulary stores and contextually salient meaning cycles distributed throughout the cortex. In turn, cycles into Wernicke's area (language comprehension), as well as the short-term/working memory area (associates object of consciousness cycles), are activated. Signs from the vocabulary and meaning cycles are comprehended in the Wernicke's and short-term/working memory areas. Next, signs from the Wernicke' s and memory areas cycle back into Broca's area for the next round of intentionality signs, resulting in the superimposition of fresh cycles upon the previous, still-reverberating cycles—forming an integrated, temporally indexed, and meaningful word stream (a sentence) for experience by the interpretant. If there is no intention to vocalize through the invocation of motor speech pathways, thinking—that is, experiencing phenomenal thoughts in words—is tantamount to listening to oneself speak, as though oneself were an audible source of words external to the experiencing microcosm. If there is intention to speak, signs from Broca's area are also transmitted into the auditory channel of motor speech pathways.

The top-down self-proficiencies of the brain derive from the causal power of its open-system holism, which, in turn, is a consequence of the extreme and self-unifying complexity of the interdependent feedback and feed-forward cycles it encompasses. In general, the complete Lagrangian governing the physical symmetry space of the brain balances the bottom-up terms which enforce the rules of local interactions, with the top-down terms which ensure the holistic integrity of the complex system. Guided by neural Darwinism, these two thoroughly intertwined, functionally reciprocative Lagrangian components seek an attractor of complexity that furthers the overall performance and viability of the microcosm.

The Lagrangian regulating brain processes is parametrized in temporal as well as spatial dimensionalities. Timing relationships among spatially distributed neuronal structures enables the harmonious collaboration and coordination necessary to achieve the brain's transcendent aptitudes. The generation of synchronicity, resonance, phasal time locking, temporal framing, and other time-dependent interrelationships, collectively called *operational synchrony*, occurs across a broad range of spatial locations and temporal frequencies in the brain. The spatial neuroanatomy of the brain has been reviewed previously, focusing on the configurations of neuronal structures and sign pathways crisscrossing among them. In the temporal domain, as measured by EEG data provided by behavioral psychology experiments,[63] interrelationships among neuronal structures have been identified. Operational Architectonics,[64] an informative statistics and modeling discipline of neuroscience, describes how standing patterns of electrical activity generated in the cortex by spatially separated neuronal structures are temporally interrelated with respect to their synchronous and resonant behaviors. Large-scale patterns of operational synchrony among neuronal structures have been correlated with sensory-limbic and executive-motor functions, cognition, information integration, and mental consciousness.

The synchronization frequency for a set of spatiotemporally organized functional structures—a neuronal group—is called its brainwave frequency. The frequency of each structure results from the aggregated firing patterns of its constituent neurons, although the emergent frequency of the neuronal structure as a whole generally differs from the individual frequencies of its component neurons. Brainwaves are produced by the interactions of interneurons and principal neurons. Interneuron complexes produce and send messages concerning the state of local firing patterns to long-range targets via principal neurons. In response, these targets return messages through principal neurons that synchronize the activity of, and generate a linkage of, brainwave oscillations.[65] Synchrony enhances the effectiveness of neural responses because correlated discharges have a stronger impact on process-structures than temporally disorganized signals; in-phase signals summate and amplify, while phase-shifted fluctuations cancel each other.[66]

Interactions among brainwaves of different frequencies are a primary factor in optimization of global operational synchrony. In general, slower rhythms provide temporal framing for higher frequency assemblies.[67, 68] Memory is another critical factor contributing to neuronal group binding through frequency selection. Based on which salience-weighted cortical memory imprints are impacting the current state of bottom-up/top-down cycling, particular brainwave frequencies are templated as selectors in *anticipation* of upcoming stimuli, and are stronger when the stimulus is more *predictable*.[69] Intentionality also plays an important role in the temporal organization of the brain. Experiments, to be reviewed in Chapter 7, involving meditators who are adept in focusing attention onto a particular object of consciousness through repetitive, intensely intentional effort, display anomalous brainwave characteristics. These experiments demonstrate that for Yogis, compared to non-meditators participating in the studies, brainwaves correlate with increased electrical power levels, sustained high-amplitude gamma-band oscillations and phase synchronies particularly in the frontal lobe, as well as widespread spatial distribution of alpha oscillations. After years of practice, high-amplitude theta-band rhythms come to pervade much of the brain. These and other brainwave modifications unique to Yogis lead to extraordinary consequences, including the performance of exceptional physical feats and novel behavioral competencies,[70] bio- and neuro-feedback control over internal physiological systems, long-term neuroplastic remodeling of the brain, and altered states of subjective experientiality.

INTEGRATED INFORMATION AND THE DYNAMIC CORE

The thalamocortical system has been of special interest to consciousness researchers, not only because the thalamic nuclei mediate the flow of all sensory signals between the external environment and the neocortex, but also because the corticocortical subset of thalamocortical cycles ascends the nodal hierarchy in which informational patterns are progressively integrated into multimodal objects of consciousness. These features of the thalamocortical system have been investigated in detail and have revealed foundational principles underlying the selection and integration of neural information relevant to the instantiation of objects of consciousness. In the companion theories[71] of *integrated information* by Giulio Tononi, and the *dynamic core* by Gerald Edelman, the thalamocortical system itself has been proposed to be the neural correlate of consciousness (although the thalamocortical system is considered by more conservative investigators to be the neural correlate of cognition[72]).

Originally, the term "dynamic core" signified the collection of all thalamocortical structures involved in the instantiation of a potential object of consciousness. Since experienceable information has now been shown to be represented in the cortex, we shall limit the term "dynamic core" to mean the collection of all cortical structures corresponding to the full neural correlate of consciousness. We also broaden the meaning of the term "integrated information" to include the qualitative semantic properties of information that characterize conscious percepts, going beyond the quantitative Shannon information mathematically regarded in the baseline integrated information theory. Further, the causal mechanisms of biosemiosis,[73] as outlined in the Introduction with respect to organic open-system symmetry spaces, will be employed in order to fully capture how the triadic biosemiotic interaction process applies to the nervous system. The biosemiotic paradigm enables identification of the physiological and functional nature of the interpretant, and the manner in which it is involved with the standing patterns of electrical activity that have been correlated with objects of consciousness instantiated in the cerebral cortex.

A fundamental feature of biosemiosis is the open-system relationship between the interpretant process-structure and the afferent and efferent sign streams with which it interfaces. In the evolved brain, the cycles looping through the corticocortical interpretant at the top of the neural hierarchy establish it both as a causal receiver and a causal producer in the process of consciousness. As receiver, afferent signs ascending to the phenomenal symmetry space are impelled by metabolic energetics, via bottom-up sensory, limbic, and cognitive pathways, for integration into a unitary object of consciousness; this is the afferent process of *apprehension*, as in hearing. As producer, efferent signs descending from the phenomenal symmetry space are propelled through top-down executive-motor and other goal-directed channels, which branch into various action pathways and reentrant feedback cycles in order to modulate attentional focus; this is the efferent process of *intention*, as in speaking.

The theories of integrated information and the dynamic core link ideas about integrated information, the process of reentry, the remembered present, and neural Darwinism (Edelman was awarded the Nobel Prize for discovering the role that Darwinian selection plays in the ontogeny and functionality of the immune system). The process of reentry refers to the synchronized, bidirectional, recursive cycling of neural pathways operating in the brain at numerous spatial and temporal scales. In particular, reentry cycles are implicated in several key facets of mental consciousness. Within the cortex, the recursive cycling of informational signs regulates foci of attention, establishes and upholds the remembered present,[74] and promotes the binding of resonant, unitary instances of integrated information. Core reentry cycles are enhanced by a profuse meshwork of interneurons that interconnects them, thereby establishing a deep-grained, horizontal level of intra-modulation.

The patterns of integrated information that actually become the objects of consciousness experienced by an interpretant are selected through the process of *neuronal group selection*[75] (neural Darwinism), which is implemented through the operational processes of the dynamic core. The criterion for the selection of a particular informational representation, from the vast assortment of possibilities available in the human cerebral symmetry space, involves an attractor of complexity. The attractor is the condition of maximum complexity practicable among afferent and reentry cycle variants produced by temporally linked neuronal structures. These variants are sampled through massively parallel loops at high frequency, refreshing within hundreds of milliseconds. Tempered by Lagrangian factors that optimize an organism's competing necessities, such as timely response to the environment or maintenance of holistic integrity, the sampled dynamic core possessing optimal complexity is the one instantiated as an object of consciousness (in the vernacular of integrated information theory: the "maximally irreducible cause-effect repertoire," φ^{Max},[76, 77] where φ signifies the amount of information generated by a system, above and beyond the information contained within the components of the system).

A mental experience is unitary; an experienced phenomenon is singular and whole. When the sight of an apple is experienced, its shape and color are always bound; we do not see these experiential qualities separately. There are opposing views regarding the unitarity of objects of mental experience.[78] The "experiential parts" school of psychologists and philosophers contends that a unified conscious experience is a composite of other experiences. The "no experiential parts" faction contends that a unitary experience is non-composite, and while a unified conscious experience may refer to a complex physical object or content, it, of itself, has no parts; particular experiences are subsumed by a single state of consciousness[79] rather than containing them as parts.

The no experiential parts theory of unitarity is supported by several general properties of neural architectures. A common characteristic of all nervous systems is that their connectomes are contiguous; there are no animals that accommodate more than one nervous system. For example, some species of hydra possess two nerve net layers, but these primitive structures are interconnected by chemically mediating synapses. Similarly, some leeches possess 32 brain centers which largely work independently, but, again, these centers are neuronally interconnected. There are no isolated neuronal islands found in any animal, and thus no clearly-cut foundations upon which multiple centers of mental consciousness are supported. Also, the large-scale integration of brain activities further indicates a unified global workspace.[80]

Another observation, based on the surgical dissection of the human corpus callosum, lends credence to the no experiential parts proposition. A severed corpus callosum (the nerve tract consisting of over 200 million fibers that connects the left and right hemispheres of the cerebral cortex) may produce bizarre effects, including the alien hand syndrome. In this experiment, a corpus callosotomy patient who consciously and intentionally reaches for an object with the right hand (controlled by the left hemisphere) has, without conscious intent, the left hand (controlled by the right hemisphere) block or undo the right-hand's attempt.

Historically, the alien hand syndrome has been taken by some to imply the existence of two competing centers of consciousness having distinct experiential parts (albeit with the qualification that only the dominant left hemisphere has the capability to later *report* its experience). But more recent findings[81] involving the interplay of the left and right premotor cortices (control of motor planning) and the left and right supplementary motor areas (mediate competing motor plans), whose functions are compromised by split-brain conditions, also explain the abnormal physiological behaviors, without the need to conjecture multiple experiential parts. Despite the unsettled experiential parts debate—whether unitarity evolved by combining parts to provide the selective advantages of consistency, coordination, and purposiveness, or whether it is a global property of the unbroken holistic interneuronal network—unitarity is a defacto feature of our conscious experiences.

SELF-AWARENESS

Some objects of consciousness represent information about ourselves. We are able to discern signs about the state of our holistic selves and distinguish them from signs pertaining to the environment. The feeling of an itch, the intention to succeed, the thought, "I'm happy," all illustrate the well-developed human aptitude of *self-referentiality*. Nuances of self-referentiality such as these are unique to human experientiality. The wide range of meaning cycles associated with human self-referentiality is a product of the extreme complexity of our integrating brains, but fundamentally, the core property of self-referentiality is intrinsic to less complex microcosms as well. Indeed, self-referentiality has its origin in the holistic closure of our autocatalytic ancestors:

> *An important aspect of system organization concerns the arrangement of processes at one or more levels. The inner process circle renews itself continuously and, as a whole, acts like a catalyst. An autopoietic [Gr. 'self-producing'] system refers in the first line to itself and is therefore also called self-referential. If a dissipative structure is forced to retreat in its evolution, it does so along the same path which it has come. This implies a primitive, holistic system memory which appears already at the level of chemical reaction systems.*
>
> — Erich Jantsch [82]

Self-referentiality provided a powerful selective advantage to evolving forms of life. Self-referential fitness, initially a holistic trait of primitive autocatalytic systems and unicellular organisms, was exported through evolution to their holistic multicellular descendants, where it has found its highest expression in the form of human mental consciousness of the self, or *self-awareness*. It is important to distinguish between subjectivity and self-awareness. Subjectivity in our usage refers to the first-person perspective that is engendered by the endogenous processes of mental consciousness, whereas self-awareness is the subjective experience of the self. The human cognitive process can both integrate self-referential "self" signs (coming from homeostatic, proprioceptive, and other holism-based lower processes of the brainstem, hypothalamus, and autonomic centers—the body), and "non-self" signs (coming from the thalamus and memory—the extrinsic world) into the remembered present of a multi-modal object of consciousness.

> *There are many forms of consciousness, such as those associated with seeing, thinking, emotion, pain, and so on. Self-consciousness—that is, the self-referential aspect of consciousness—is probably a special case of consciousness.*
>
> — Francis Crick and Christof Koch [83]

Language is a cognitive ability that facilitates both internal thinking and external communication through verbal symbols. Verbal symbols can recruit or be recruited by both non-self- and self-referential meaning cycles. We identify the particular experiences of a human interpretant that incorporate self-referential meaning cycles as self-awareness. The linguistic token "I," for example, recruits self-referential meaning cycles encoded in the cortical memory store, which can be fetched by the working memory process to be employed as a component of integrated information; thereby associating self-referential meaning cycles and the symbol "I" into an object of consciousness, and therefore self-awareness of the interpretant. Conversely, the symbol "I" may be recruited by meaning cycles, as when answering the question "How are you?" with "I' am well." Depending on prevailing context, however, self-awareness may or may not be instantiated into an object of consciousness. "Seven"; note that no self-referential cycles were integrated into the reader's object of consciousness when the number seven was experienced. But when self-referential cycles and meanings *are* cognized and integrated into an object of consciousness, there arises the experience of self-awareness, "I" am seven years old.

MEDITATION

The practice of yoga was born in the ancient East. In its earliest form, yoga involved the *intentional* control of the bodily functions, breathing, and the mind. The Hindu Upanishads, written in Sanskrit 3,000 years ago, refer to the practices of *pranayama*—the intentional control of the breath, and *pratyahara*—the intentional attenuation of sensory channels. As time progressed, meditational techniques concerned with intentional control over cerebral processes were developed and added into the yogic system.

Various experiments in contemplative neuroscience have been conducted on well-practiced meditators to measure the effects of meditation on their brain function and physiology. In a cooperative study involving more than 20 universities over a 15 year period, a group of Buddhist monks, each having more than 10,000 hours of experience in meditation, participated in this testing project, some at the behest of the Dalai Lama. The results indicate that the experienced meditators (as compared to norm groups also tested) possess extraordinary abilities to control their bodies and brains, and that the structural anatomies of their brains actually transform over time. The long-term effects of meditation on the brain were also found to depend on the type of meditation practiced. In the experiments involving the Buddhist monks, three traditional meditation techniques were analyzed: one-pointedness, mindfulness meditation, and compassion meditation.

One-pointedness is the original technique of meditation, dating back millennia to the earliest yogic customs. The practice of one-pointedness is the holding of a single object of consciousness in focus for an extended period of time. Initially, objects of consciousness consisted of morphemes (minimal grammatical units of a language), but evolved into composite words and verses. To achieve one-pointedness requires that all unrelated cycles, including extraneous sensory signals, visceral feelings, breathing, recursive thinking, etc., be intentionally inhibited from integration into an object of consciousness. The practice of one-pointedness develops a person's ability to respond quickly to the changing situations that arise in the course of daily events, and the ability to focus attention intensely for prolonged periods of time.

Sanskrit symbol for the sacred sound, "Om."

Om is an ancient mantra, composed of three phonemes, "a-u-m." Om signifies the cosmic breath, the essence of ultimate reality, and consciousness.

In mindfulness meditation, informational cycles are intentionally welcomed into a continuously changing flow of consciousness. However, the particular cycles associating excessive emotion and worldly attachments into consciousness are reflectively inhibited. This practice develops one's ability to cope with distressful thoughts and traumatic events.

Compassion meditation requires continued effort to cultivate awareness of the suffering of others and to intentionally experience sincere concern and empathy. These meditative practices employed to construct chosen phenomenal symmetry spaces, through the prolonged intentional regulation of meaning cycles, have proven to be of great value, both personally and in dealing with society. Through mindfulness meditation, disturbing and unhealthy personal emotional cycles may be reduced or removed; while in compassion meditation, meaning cycles signifying sympathy and helpfulness toward others are reinforced.

Expert meditators are able to manipulate the locations and amplitudes of electromagnetic activities throughout their nervous systems. Through neurofeedback, they can intentionally take control over their autonomic nervous systems and associated bodily functions in a manner distinct from those evincible by non-adepts, as shown not only by their physical feats and purity of being, but by empirical experiments demonstrating that yogis can intentionally control electrical activities throughout their nervous systems, as when they generate ultra-high-amplitude electrical spikes in the palms of their hands by means of techniques called *mudras.* In the brain, one-pointedness and mindfulness meditation have been correlated with various electromagnetic anomalies; including sustained high-amplitude gamma-band oscillations and phase synchrony in the range between 25 and 42 Hz, particularly in the frontal lobe; as well as increased power levels and spatial distribution of alpha oscillations. After years of practice, high-amplitude theta-band rhythms come to spread over much of the brain.

The neuroimaging experiments demonstrate that long-term meditation causes regions of the brain to change in size. In general, the prefrontal and insular cortices, involved in attentional and executive functions, increase in thickness for meditators compared to control groups, while specialized meditative techniques correlate with specific neuroanatomical size alterations. In the practice of mindfulness meditation, the persistent inhibition of emotional attachments results in decreased amygdalar volume (to be anticipated on the basis of the strong correlation between emotional behavior and amygdalar activity). Other adaptations to brain structure resulting from long-term meditation include changes in neural pathway densities and remapping of interfunctional networks. Thus, the neuroplasticity of the brain allows meditators to intentionally mold their own interpretant process-structures and to experience complementary altered states of consciousness.

Chapter 7

THE EXPERIENTIAL ESSENCE OF MENTAL CONSCIOUSNESS [i]

Describing the essence of mental experience, or even the phenomenal content—the qualia—of a mental experience poses difficulty, because experiences and their qualia are exclusively intrapersonal and thus ineffable. An oft-used depiction to portray the subjective nature of qualia is: *Qualia is an unfamiliar term for something that could not be more familiar to each of us: the ways things seem to us.*[84] Another frequently cited reference to the essence of experience and qualia is: *What is it like to be a bat?* [85] What things are *like,* or *seem* to be, are not subjects of the hard sciences. Nevertheless, based on consensus, there are general characteristics of human experiences and their qualia about which there is agreement. They are private (isolated to an individual phenomenal symmetry space), subjective (first-person perspective), non-relational (red is independent of the object it qualifies), and unitary.

The experiential essence of mental consciousness is an emergent property that supervenes upon the collective sensation of integrated information. It emerged with the bifurcation and re-association of interneurons and sensory neurons in primitive nervous systems, thereby establishing the

---

i *Experience*: the direct participation in or the observation of events as a basis of knowledge.
Phenomena: (in the sense of phenomenological psychology; subjective experience) things as they appear to and are constructed by the mind (the domain of mental consciousness); as opposed to *Noumena*: things which in themselves are inaccessible to the mind and to which phenomena are referred.
[...*humans could infer only as much as their senses allowed, but not experience the actual object itself.* —Kant]
Mental consciousness: the faculty of the mind by which it experiences phenomena.
Essence: the intrinsic nature or indispensable quality of a thing. The *essence* of mental consciousness is independent of phenomenal content.

interneuronal domain. Interneurons and principal neurons in the interneuronal domain exapted the genetic bases of unicellular sensation, by which they collaboratively coevolved the self-proficiency and the phenomenal essence of collective sensation, that is, mental consciousness.

In the human brain, where evolution has attained the highest degree of complexity we have yet discovered in the universe, the process-structures that underlie mental consciousness perform the necessary and sufficient functions to engender the vast range and richness of our experiences. The human brain employs an intricate web of interdependent and unifying functions: prolonged and stable cycles of ultrasensation, spatiotemporal segregation of the informational representations selected for the phenomenal symmetry space, an endogenous orientation that enables a first-person, subjective perspective on those representations, and the powers of purposeful intention.

Human mental consciousness marks a plateau of evolution. Indeed, to the best of our knowledge, evolution's grand achievement is our very own vehicle of consciousness—the human brain.

> *Man carries the world in his head, the whole astronomy and chemistry suspended in a thought. Because the history of nature is charactered in his brain, therefore he is the prophet and discoverer of her secrets.*
>
> — Ralph Waldo Emmerson

* * *

8
Consciousness Unfolded

The dimensionality of the human brain's phenomenal symmetry space is extremely high, reflecting the many qualia and threads of meaning it integrates into a unitary object of consciousness. Based upon correlations observed in cognitive psychology and neuroimaging experiments, the blending of multiple qualia and abstract meanings—signifying high-level, multi-modal instances of integrated information— into precise, but diversified, objects of consciousness depends upon the spatiotemporal infoldment of brainwaves it assembles, through complex electromagnetic interactions, into synchronous and resonant patterns. However, all other living beings, regardless of complexity, are able to integrate information, albeit with a lesser degree of dimensionality, and therefore they can only produce objects of elementary consciousness; or, in the case of organisms with less developed inter-neuronal domains, simple objects of mental consciousness. In general, all living systems engage their own particular forms of process-structure and knowledge-base to integrate the information that determines the type and quality of their experientiality.

Taking the integrated information generated by any living system to be a correlate of elementary or mental consciousness, permits generalization of integrated information theory. To date, integrated information theory has been applied primarily to the complex neural networks infolded within the thalamocortical systems of higher vertebrae; encompassing the working memory subsystem located in the prefrontal cortex at the top of the integration hierarchy, and addressing only the quantitative aspects of information in terms of the Shannon measure. We will extend the basic ideas of integrated information theory to include less complex, less infolded—*unfolded*—neural and non-neural microcosmic systems.

Consider a *frog,* a neural organism having neither a prefrontal-cortex-based working memory system, nor even a developed neocortex, but yet is conscious. In this case the semiotic interpretant process-structure is based in the three-layered allocortex where it is seen that operational modules produce standing patterns of electromagnetic activity. These electromagnetic patterns have been empirically correlated with a frog's conscious behavior. A frog also has a thalamo-basal ganglia-allocortical cycle, which enables it to keenly apprehend sensory signals and convert them into intentional motor responses. But instead of being integrated with support from a multimodal prefrontal cortex and its working memory selection system, after a few synaptic steps in the frog's heteromodal cortices, only here-and-now sensations and immediate needs are integrated.[86] Without the multimodal, and rich-in-meaning, symbol generator of the infolded working-memory process-structure in the prefrontal cortex, the frog is equipped to execute only relatively simple cognition cycles, and is limited to experience only instinctual objects of consciousness in its phenomenal symmetry space.

A *dolphin* has a brain structure that is intermediate between the frog and the human being in terms of infolded complexity. Whereas the frog lacks a developed neocortex, the dolphin inherited one from the land mammal *mesonyx,* which it carried it back into the sea 50 million years ago. With intricately organized temporal and parietal neocortices, the dolphin is able to exhibit emotions, altruism, mourning, cognitive intelligence, and self-awareness.[87] However, the frontal lobe of the dolphin's neocortex is relatively undeveloped, and thus the advanced integrative capabilities associated with the human dorsolateral prefrontal cortex where working memory is centered, are absent. The dorsolateral prefrontal cortex, one of the most recently evolved parts of the human brain, is both the terminus of the sensory stream and the origin of the executive-motor pathway in the nodal hierarchy, as well being at the causal cycling nexus of highest-order, multimodal, abstract informational representations, well beyond the capacities afforded by the dolphin's temporal and parietal cortices.

The worm *Caenorhabditis elegans* was the first multicellular organism to have its entire genome sequenced. It has 302 neurons,[88] including six nerve cords that run the length of its body, with each cord projecting neurons that innervate organ tissue.[89] The nerve cords are joined together by a nerve ring, called the cerebral ganglion, which forms an elementary proto-brain where input signals from chemical and pressure sensitive neuronal receptors are integrated. Conversely, the cerebral ganglion is an executive center from which signals are disseminated to control its muscle and secretory organs (secretion of hormones into the extra-cellular space is a pre-neuronal chemical mechanism that modulates global salience levels by activating or inhibiting functional systems).

The ebbs and flows of the material transport systems that integrate the functional cycles of a non-neuronal microcosm, such as a *tree*, are certainly far-removed from the electromagnetic signals whizzing around in the human brain. But the underlying mechanisms are analogous. In a tree, infoldment is relatively low, and objects of consciousness are instantiated in spatially dispersed, temporally asynchronous cycles. For example, a root-based water assimilation system and a leaf-based energy photo-synthesis system are integrated in a resource transport cycle modulated by hormones traveling through vascular pathways: water and nutrients to the leaves via xylem ducts, and energy-bearing sugar to the roots via phloem ducts.[90] Thus, signs from the environment such as lack of water during a time of drought are converted into meaningful information relative to the tree in its aspect as holistic interpretant, inducing it to project roots to probe for water. Reflecting the global effects of the slow-moving hormonal agents integrating the systems of a tree, corresponding objects of elementary consciousness are trivial and blurred in space and time, well below the informational dimensionality of a neural symmetry space and thus the threshold of human apprehension. But even though the interpretant process-structure of a tree is dimensionally unfolded, it still possesses and displays holistic properties that are shaped by its genotypic characters, memory, global hormonal contextualization,[91] and the semantic nature of the information it processes.

A *unicellular organism* is a holistic microcosm that can interface with its environment. Primitively structured unicellular organisms rely on specialized RNA/DNA-based coding centers that are widely dispersed throughout the cytoplasm: the mitochondria for metabolism, micronuclei for reproduction, macronuclei and ribosomes for biosynthesis, etc. These organelles and nuclei interact via diffuse chemical symbol transfer cycles, therefore a sequestered semiotic interpretant process-structure is not formed; there is no hardwiring to facilitate the temporal infoldment of integrated information, and thus objects of elementary consciousness are undeveloped. However, it is interesting to speculate about the consciousness of a unicellular organism, entertaining the idea that it is not so trivial after all. Certain species of eukaryota can perform various actions that demonstrate the faculties of apprehension and intention: for example, the amoeba *Naegleria gruberi* can seek-out, detect, and recognize signs from paramecia, resulting in their engulfment and assimilation.

It is conceivable that there exists a process-structure infolded within the DNA complex of a unicellular eukaryote that supports a form of mental consciousness analogous to the neural brain. The parallels between the symmetry spaces of the human brain and the DNA complex of an amoeba are extensive. Both the brain and the DNA complex are open, biosemiotic systems with a symmetry space having a holistic Lagrangian factor. Also, they are both highly complex integrators of information. The emergence of a phenomenal symmetry space in a DNA complex would require that it embody an interpretant process-structure with an embedded plastic memory, which the DNA complex does have. DNA transcription centers are hubs of complexity where afferent signs (transcription factors) and efferent signs (messenger RNAs) are processed to facilitate gene expression. Analogously, this spatially infolded, selective, "working memory" processor could be a focal point for a variable "symbol," hard-wired to the experiencing DNA interpretant conformation in which memories are embedded. The structure of the DNA interpretant could be the contextually malleable DNA/crystal-histone network, whose conformational stresses in global structure, as well as in local DNA-windings, are known to induce transcription process selection. Could this interpretant integrate multimodal transcription influences and experience them? There still remains much to be learned before judging the feasibility of a unicellular organism being non-trivially, proto-mentally, conscious. Regardless of these speculations, the fundamental processes that begat human consciousness are innate to all life forms, although modes of experientiality vary greatly with platform and degree of infoldment. Thus, for a frog, a tree, or a cell, there is no reason to anthropomorphize their conscious experiences by ascribing human qualia to their objects of consciousness.

In addition to providing an explanation for the origin and continuous evolution of consciousness along the anthropic pathway, the broader, unfolded theory presented here offers epistemic advantages as well. It reinforces the age-old intuition that consciousness is an intrinsic property, or potentiality of reality. Throughout history many philosophies and religions have been built upon the premise that consciousness is the very essence of reality; or, at least, that it is separate from but dual to the material world. Our broader view allows us to establish a general theory that incorporates the scientific finding, described in Chapter 2, that the evolution of complexity and integrated information has proceeded continuously from the low-entropy potential for order inherent in the homogeneity of the big bang universe.

A model that includes unfolded varieties of consciousness also obviates restrictions which had heretofore precluded a coherent explanation for how, when, and why consciousness arose in the first place. In this regard, the unfolded view is more unifying than exclusively infolded theories which, for example, sets the baseline for the existence of consciousness to be exclusively that state of the brain where maximally compacted neuronal circuitries vibrate at maximally compressed cycle times of 1/40 second. Rather, the ascent of consciousness to the caliber attained by the human brain has entailed a series of evolutionary phase transitions in the microcosmic symmetry space, superimposing layers of informational abstraction in a dimensionally holographic hierarchy. In particular, key informational phase transitions occurred at: the autocatalytic closure of abiogenic chemical networks into holistic unicellular organisms, the holistic closure of colonies of unicellular organisms into multicellular organisms, and the closure of neural transport networks into integrated, synchronous cycles of the brain.

As a thought experiment, if a human being's neural architecture were unfolded into a slower, asynchronous and spatially dispersed network of information processing mechanisms, but retained (diminished) holistic interdependencies, it would resemble, for instance, that of a sea star. Likewise, if the dynamically-congregating, function-integrating chromatin loops in the DNA hub of a eukaryotic cell were unfolded into a stoichiometric autocatalytic chemical mixture, it would be similar to a vent-chambered proto-organism. What is common to all of these life forms is that they are holistic, open-system producers of meaningful integrated information. Hence, amoebae, trees, frogs, and people all produce objects of consciousness, elementary or mental.

Consciousness did not suddenly appear in the brains of highly evolved animals on earth. It is the product of an evolutionary process that has sought an attractor of complexity, and resulted in the formation of holistic microcosms endowed with a self-embodied knowledge base and ability to integrate information. All living systems apply their knowledge base to the information they integrate, thereby characterizing their encounters, i.e., their experiences, in the world. The nature of experiences varies widely: from the unfolded, unmindful interactions distinguishing the elementary consciousness of an amoeba or a tree; to the highly-infolded, subjectively-perceived objects of mental consciousness realized in the human

9

Other Perspectives

This chapter supplements the science-based description of the evolution and nature of consciousness that is the principal subject of this book. To consider other perspectives on the understanding of consciousness, we will explore those offered by philosophy, psychology, and theology. In particular, the ancient Eastern system of the tattvas will serve as a backdrop for the other approaches, because the tattvas describe the realities of consciousness in terms of insights ascertained directly through turning inward to the personal world of one's own experientiality.

THE MICROCOSM

Researchers concerned with the subject of consciousness, especially those engaged in the hard sciences, have eschewed experience-based "spiritual" systems of the East, steering clear of the occult and superstitious baggage they have accrued over the centuries. But, in the wake of neuroscientific evidence that has come to light in recent years regarding the practice of meditation and its effects, it is important to consider what the masters have to say about the experiences they instantiate in their self-structured brains.

The philosophical essence of the early Eastern discoveries regarding the nature of consciousness was described in the ancient Sanskrit system of *tattvas* (Skt. principles). We will commission certain terminology and ideas from these teachings in order to incorporate the nuances of experientiality they provide. The following account of the tattvic system omits deistic and ritualistic connotations, and specifically addresses the hierarchical symmetry structure of the secular core system, the intrinsic process of consciousness, and the supervening property of experientiality. We will build upon and modernize the account of the Shiva-Shakti Tattva as it is presented in *The Garland of Letters* by Sir John Woodroffe.[92]

The ancient wisdom naturally impacts our understanding of personal responsibility and cosmic purpose. While the scientific accomplishments of our time are important in their own right, having fostered progress in consciousness studies, engineering, medicine, and so forth, they are not directly instructive when it comes to achieving optimal conduct among persons and societies. Scientific advances in genetic engineering, nuclear energy, and the awareness of global warming have introduced ethical uncertainties. Thus, intimate knowledge about how one's own process of consciousness may be controlled and shaped is of value, not just for an academic understanding of consciousness, but for acquiring a vehicle of consciousness through intentional regulation of experiences which may optimize personal fulfillment in a world of complication and challenge. In the modern-day science-oriented culture of the West, each generation witnesses a resurgence of interest in the old teachings of the spiritual East. This is because of the persistent feeling that a profound truth or path to self-improvement lies within these enduring consciousness-based beliefs, traditions, and practices.

The original Sanskrit writings employed a logical, hierarchical, syntactical structure. The hierarchical organization of the human symmetry space was first discovered by the ancient yogis who traveled inward, to the depths of personal experience. The earliest doctrines pertained to the inward analysis of the experiential states resulting from meditation on a single object of consciousness, such as one-pointed concentration on, and repetitive mental recitation of, a single phonetic sound (morpheme)—a primal *mantra*. The yogis assigned meanings to these (perceived) vibrational sounds, and later to their composite words and verses. The resulting vocabulary of experience-based concepts that arose from meditation on mantras, supplemented by other yogic techniques, have no close counterparts in modern languages. However, these "primitive" concepts have reemerged in state-of-the-art fields of science; hence, the vocabulary and insights provided by the tattvic system serve to be practical and unifying.

Based on techniques of self-purification, self-control, and introspection, the ancient seers formulated a model of the environment, body, mind, and consciousness. Their discoveries established the basis of the Samkhya philosophy documented in the oldest Sanskrit scriptures, which enumerate a group of 25 tattvic principles arranged in a hierarchy ascending from the external environment, through the internal systems of the body, to the seat of consciousness.

The 25 tattvas are partitioned into five categories based on their physical subtlety, corresponding to the degrees of integration and abstraction of the sign types that characterize the hierarchy; see Table 9-1 (note that tattvas 33 and 34 are interchanged to correspond to the density gradation of modern physics). In ascending order, the tattvic categories designate the *bhutas*—the gross elements, the *tanmatras*—the subtle elements conveying signs from the environment to the microcosm, the *jnanendriyas*—the afferent sensory organs conveying signs to the interpretant process-structure in the brain, the *karmendriyas*—the efferent motor organs conveying signs from the interpretant process-structure, and the *antakaranas*—the three levels of the mind that correspond to the cortical cycles of the interpretant. Together, these tattvas constitute the material world of *Prakrti.* The highest and subtlest level in the regime of Prakrti is the *buddhi* tattva. In neuroscience, the buddhi is equivalent to the electromagnetic fields of the brain which signify the integrated information embodied in the conjunction of signs and the semiotic cortical cycles of the interpretant. Situated above the material realm of Prakrti is the unique and supreme tattva of *Purusa,* the non-material experiential faculty that supervenes upon the phenomenal symmetry space.

Antakaranas		*Internal organs*
12	*Purusa*	nonmaterial consciousness
13	*Prakriti*	material potentiality
14	*buddhi*	intellect
15	*ahamkara*	ego
16	*manas*	lower mind
Jnanendriyas		*Sensory organs*
17	*srotra*	audition
18	*tvak*	tactile
19	*caksu*	vision
20	*rasana*	taste
21	*ghrana*	smell
Karmendriyas		*Motor organs*
22	*vak*	speech
23	*pani*	hand
24	*pada*	foot
25	*upastha*	evacuation
26	*payu*	procreation
Tanmatras		*Subtle elements*
27	*sabda*	sound
28	*sparsa*	touch
29	*rupa*	form
30	*rasa*	taste
31	*gandha*	olfaction
Bhutas		*Gross elements*
32	*akasa*	space
34	*tejas*	radiation *(agni)*
33	*vayu*	gaseous state
35	*jala*	liquid state
36	*prthyi*	solid state

Table 9-1 *The Microcosm*

What is significant about the Purusa-Prakrti architecture is how closely it is resembled by the modern multi-discipline naturalist model, which consolidates neuroanatomy, psychology, and other applications of the scientific method. Given that this ancient civilization had virtually no scientific technology, the similarity between the tattvic and scientific systems strongly suggests that the yogis must have had some other form of intimate functional contact with the nervous system. For example, both the physical scientific and the experiential tattvic architectures depict a linear pathway that connects discrete functional centers—the neural tube connecting ganglia, and the *susumna* connecting *chakras.* The linear pathway, as described in both of these systems, is bi-directional—the afferent and efferent nervous system channels carrying signals to and from the brain, and the *ida* and *pingala nadis* (channels) conveying signals to and from the mental antakaranas.

Both the scientific- and experiential-based models describe bottom-up integration of functionality and abstraction along their linear pathways, culminating as objects of consciousness in the seat of phenomenal experience—the centrosemiotic interpretant in its subjective aspect, and the Purusa. Correlatively, both systems postulate top-down causality associated with the Purusa. At this point, adopting terminology reflecting the top-down Purusa perspective becomes constructive to our inquiry, as it refers the discussion of consciousness to the first-person vantage point of the masters, who have expertise in controlling, shaping, navigating, and experiencing the dimensionalities of their phenomenal symmetry spaces, as demonstrated in the experiments involving meditation.

Since the developers of the tattvic system experienced the antakaranas as being intentionally causal to the gross physical levels, these tattvas were regarded as the underlying informational source upon which the very existence of the lower instruments depends. This would seem to be in conflict with modern, bottom-up, reductionist theories of physiological evolution, according to which the subtlest levels of electromagnetic focus and integration, in the form of brain waves, have been the most recent to evolve. But with respect to the evolution of self-organizing microcosms—particularly as vehicles of consciousness—the intentional assimilation, biosynthesis, and mental control of informational resources reflects the generative causality intrinsic to the holistic microcosm. Whereas in neurophysiology, the seat of human consciousness is associated with the centro-semiotic interpretant at the top of the cephalization hierarchy of physical vehicles, similarly, according to yogic terminology, Purusa is at the top of the experiential hierarchy, which is also correlated with physical vehicles, called *kosas.*

The kosas are the material organs of the human microcosm that embody the tattvic hierarchy. Thus, the kosas ascend in subtlety from the non-neural organs, through the nervous system and the levels of the mind, to the *anandamaya* kosa, the abode of the buddhi tattva. The sutras of Patanjali (400 AD) detail *raja yoga,* "the royal road," which includes eight specific practices by which all of the kosas may be purified and balanced *(sattva),* with the ultimate goal that their influences be dissolved and precluded from intruding into *pratyaya* (objects of consciousness). These practices enable the aspirant to achieve degrees of both physical and experiential transformation, and thereby realize the truths put forth by the sutras.

The first two practices of raja yoga, *yama* and *niyama,* concern one's behavior with respect to the external world, while the other six lay out techniques for internal development. Yama and niyama specify a code of precepts that leads to the fulfillment of one's *dharma*—the moral obligation to act in accordance with the natural order. Yama and niyama must be mastered before sattva buddhi and inward realization can be achieved. A few of the precious gems of wisdom found in the doctrine of yama and niyama may be paraphrased as follows:

Harmlessness toward all creatures, great and small.

Meet misery with compassion, virtue with gladness, and vice with indifference.

As improper thoughts, emotions, and actions cause endless pain and ignorance, there is the necessity of pondering over the opposites.

The six techniques following yama and niyama are directly concerned with the internal symmetry space of the microcosm, namely: *asana* (the familiar hatha yoga of bodily postures), *pranayama* (control of the breath), *pratyahara* (inhibition of neuronal sensory impressions), *dharana* (mental concentration on a single pratyaya), *dhyana* (inhibition of self-awareness into the pratyaya—recall that self-awareness connotes holism-based, self-referential neuronal cycles), and *samadhi* (merger of the experiencer and the experienced, i.e., the union of signs and interpretant).

The experiments (described in Chapter 7), which employed advanced neuroimaging technologies, have established that yogis have command over their gross bodily functions, and can regulate the electrical fields of their nervous systems in extraordinary ways. These tests demonstrate their abilities to sustain high amplitude spikes at the hand plexus through the technique of *jnana mudra* (in conjunction with pranayama),[93] and to generate atypical brainwave patterns in distinctive frequency ranges during meditation.[94] Also, they are able to alter the component sizes and interconnectivities of their neuroanatomical process-structures through prolonged, intentional meditation praxes.

One who has put forth even a moderate effort to take advantage of these well-documented techniques of meditation generally recognizes a subtle transformation of body and mind. The novice begins to notice improved health, emotional equilibrium, and increased mental acuity. One who has achieved competency in the practice of *samyama* (mastery over higher mental functions) develops the intuition and wisdom of the buddhi, and, in the highest stage of *samadhi*, the union of experiencer and experienced is realized.[95]

> *Although thus quiescent there is cognition of the mental motion of thoughts arising and vanishing ... one recognizeth the 'Moving' and 'Non-Moving' as being one and inseparable. One realizeth the 'All Discriminating Wisdom'... the wisdom of the illuminated mind of the buddhi, whence comes the divine power of knowing each thing separately, yet all things at once.*
>
> — Tibetan Yoga and Secret Doctrines [96]

Patanjali's second sutra defines yoga: *Yogas citta-vritti nirodha,* "Yoga is the inhibition of the modifications of the mind." The neuroimaging experiments confirm that during meditation, when the vibrations of the mind *(citta-vrittis)* have been intentionally put into cessation (*nirodha*), yogis eliminate gamma-band frequencies from their brain waves, whilst anomalous theta-band frequencies arise. Remarkably, the masters report that the experiential states they achieve are *not* empty, as in non-REM sleep where gamma frequencies are also vacant; rather, their experiences have novel characteristics. Yogis report that as the kosas are successively transcended, their sense of subjectivity wanes—the experiencer and the experienced merge into oneness, consistent with the teachings of ancient doctrine.

The sutras further describe realms of consciousness that are forever beyond the reach of scientific method. However, regarding fundamental symmetry principles, it is important to note that the sutras culminate with transcendence into an unbroken symmetry space. According to the more esoteric sutras, *saguna samadhi* (having qualities) involves physical properties, described above, in that there is a specific (physical) pratyaya into which the experiencer and the experienced merge. But the sutras go on to describe *nirguna samadhi* (without qualities, qualia, or meanings; transcending the specific informational representations of the phenomenal symmetry space), where the seeds of all attachments to Prakrti *(samskaras)* are shed, and enlightenment, or *Kaivalia,* ensues. Similar scenarios are posited in other Eastern belief systems as well. As legend holds it, the rare few who do attain personal enlightenment may intentionally reenter into the world of attachments, for the good of all.

* * *

In the ancient West, Plato and Aristotle believed the human mind to be distinct from the physical body. In the 17th century, Rene Descartes reinforced this notion with his "Cartesian duality." In his famous maxim: *Cogito ergo sum: "I* think, therefore *I* am", Descartes identified the thinker, *"I,"* to be an unchanging non-material mind, or soul, that interacts with the material world through experienced thoughts. One may doubt the reality of the material world, including the body, but this doubt is itself a thought, and therefore proves the primacy of the thinker. In contrast to this view, which asserts the primacy of the non-material thinker, a more recent class of theories, motivated by discoveries in neuroscience and genetic genealogy, contends that it is the thought, a material incidence, which is primary, while the thinker is secondary, even illusory:

Question: What is the relation between the thinker and his thought?

Answer: Now, is there any such relation, or is there only one thing, which is thought, and not the thinker? Because, if there are no thoughts, there is no thinker. When you are thinking, when you have thoughts, is there a thinker? If you have no thoughts at all, where is the thinker? ...There is first the process of thought, and then thought creates the thinker.

— Jiddu Krishnamurti [97]

In Krishnamurti's analysis, "the thinker is the outcome of thought" refers to the thinker as being the phenomenological product of the process of consciousness, secondary to the physical machinery of the brain, which is the constructor of noumenal thoughts.

Along this line of reasoning we also have:

Each pulse of cognitive consciousness, each Thought, dies away and is replaced by another. The other knows its own predecessor, and finding it 'warm', greets it saying: 'Thou art mine, and part of the same self with me.' Each later Thought, knowing and including thus the thoughts which went before, is the final receptacle—and appropriating them is the final owner—of all that they contain and own. Each Thought is thus born an owner, and dies owned, transmitting whatever it realized as its Self to its own later proprietor... Its appropriations are [however] less to itself than to the most intimately felt part of its present Object, the body, and the central adjustments, which accompany the act of thinking, in the head. These are the real nucleus of our personal identity...The passing Thought then seems to be the Thinker; and though there may be another non-phenomenal Thinker behind that, so far we do not seem to need him to express the facts.

— William James [98]

When James states, "and though there may be another non-phenomenal Thinker behind that, so far we do not seem to need him to express the facts," he acknowledges the possible existence of a non-phenomenal witnessing soul, or homunculus, but it is not *necessarily* required for the perception of a thought. Today, the primacy of thinker vs. thought is a contentious issue, but in the sciences, the opinion held by Krishnamurti and James—that the thought alone is a sufficient and inclusive basis to explain an experiencing thinker—has gained favor. Expressed strongly by Gerald Edelman: "The temptation to appeal to a witness, to a homuncular self, must be resisted." [99]

The temptation to appeal to a witnessing self is hard to resist, because of the ethereal feeling of abstract thought, and the persistent intuition that experience involves more than just a series of physical patterns appropriated by a complex electrochemical systems. The solution to the primacy dilemma of thought vs. thinker—or seen vs. seer, or material vs. experiential, etc.—is that all of these standpoints have validity and are complementary aspects of an inclusive ontology. A physical object of consciousness that is represented in the phenomenal symmetry space of the human microcosm (the thought), by the process of consciousness (thinking), is experienced by a collective interpretant structure (the thinker). The thought refers to the informational representation instantiated by the interaction between a sustained, reentrant signal pattern and the collective structure, while the thinker is the collective structure itself, which has a first-person—self—relationship with the information it both conjoins and embodies. The thinker has the faculty to collectively sense, to experience, the information represented in its phenomenal symmetry space.

* * *

THE MACROCOSM

Shiva-Saktyatmakam visvam.
"Shiva and Shakti are the innermost essence of the universe."

This ancient Sanskrit tattva is the principle that introduces the subtle relationship between consciousness—Shiva, and the macrocosmic power by which it becomes manifest—Shakti. The Shiva-Shakti Tattva does not refer to a duality between mental and physical worlds directly, but rather to the guiding principle for the evolution of the universe, including the realization of conscious experience. The Shiva-Shakti Tattva describes the process by which physical manifestation and conscious experience emerge within *Brahman*, the permanent condition of balance that unifies all. Within Brahman's symmetry, however, there exists a fundamental instability arising from the nature of the sub-symmetries it enfolds.

In scientific terms, the manifest universe is the effect of the breaking of the unstable (tachyon) supersymmetry that existed at the big bang. Yet, while this symmetry principle is common to both the tattvic and scientific cosmogonies, there is an important distinction essential to the formulation of a complete theory of consciousness. Whereas modern scientific systems are more detailed and mathematically developed, the Shiva-Shakti Tattva expounds a conceptual framework that incorporates the phenomenal dimensionalities of conscious experience.

By extrapolating the characteristics of Purusa consciousness, eleven more tattvas, beginning with *Shiva* and *Shakti,* were appended to the chart in the Shaivite philosophy in order to expand the tattvic system from local microcosms and incorporate the macrocosmic aspects of reality (see Table 9-2). The new tattvas were attributed as *potentialities* of Shiva—the universal substratum *of* consciousness. Inherent to Shiva, is Shakti (Skt. stress), the power to manifest and animate the *Dance of Shiva, for* consciousness. The Shiva and Shakti principles have been highly anthropomorphized and deified in Eastern belief systems and traditions over the ages, but when consciousness is considered as a cosmic process principle leading inevitably to the emergence of human experience, in modern scientific terms, the power of Shakti is analogous to the gauge-forces (see Chapter 13) unfolding in the low entropy plenum of potential order, complexity, and consciousness, which is Shiva.

As with the Purusa-Prakrti Tattva of the microcosm, the Shiva-Shakti Tattva closely resembles the scientific description of the macrocosm. The Shiva-Shakti Tattva describes cosmogony with concepts corresponding to hierarchical symmetry breaking *(kancucas)* and causal information *(vidya);* foundations of theoretical physics. Cycles of the universe are said to emanate from an overarching state of complete but inherently unstable symmetry (Brahman), into the plenum of conscious potentiality, Shiva. Technically, the term "Shiva-Shakti Tattva" denotes only the *energetic* aspect of manifestation, but is used generically to encompass its dual aspect, the "Maheshvara-Maheshvari Tattva," which is concerned with the *informational* properties of nature. [100]

Suddha Tattvas		
1	*Shiva*	consciousness – principle in substratum
2	*Shakti*	stress, global potential – makes Shiva conscious
3	*saddashiva*	power of will – subjectivity, 'I-ness'
4	*ishvara*	power of knowledge – objectivity, 'this-ness'
5	*suddha vidya*	power of action/pure knowledge, I-this balance
6	*suddha maya*	impure knowledge – limitations on I and this
Kancucas		
6	*maya*	illusion of individuality, duality (*ma* – 'finite')
7	*kaala*	power limitation
8	*vidya*	knowledge limitation
9	*raga*	attraction, motion limitation (*dvesa* – 'repulsion')
10	*kalaa*	temporal limitation
11	*niyati*	spatial limitation
12	*Purusa*	limited consciousness

Table 9-2 *The Macrocosm*

In every cosmic cycle, the Shiva-Shakti and the Maheshvara-Maheshvari tattvas emerge together from the even higher symmetry principle of *Paramashiva,* the "source of all." Described scientifically, Paramashiva is the basis for the universe to arise from the condition of maximum potential energy through Shiva-Shakti, and maximum informational potentiality—inherent in the minimal gravitational entropy of the big bang—through Maheshvara-Maheshvari.

Although there is strong correspondence between their energy- and information-based symmetry principles, the tattvas should not be taken merely as vague progenitors of modern physics. The tattvas encompass the undeniable reality of inner experience which has been beyond the reach of science, until now. Of course, while the properties of experience discovered by the ancients have garnered strong support by inference from scientific evidence, actualized qualities of experience can only be *known* within, and by, a microcosm. Thus, the tattvic model unveiled by probing the inner depths of experience through practices such as meditation sheds a fuller light on the character of consciousness.

The nature of the underlying essence of the universe, physical manifestation, and consciousness, have long been debated in both Eastern and Western cultures, with schools of thought spanning a wide spectrum of ideas. In the East, the extremist *vivarta-vada* philosophy long ago asserted that only consciousness is fundamentally real, while physical and mental manifestations are illusory, because only consciousness is permanent and unchanging. In the Eastern counterpoint view of *abhasa-vada,* the status of reality is attributed to phenomena; these being the projections of an underlying consciousness-centric ultimate reality. In the West, the notion of an ultimate reality was developed as well; for example, in Plato's mathematically perfect "ideal forms." But Plato explicitly denied the realism of the visible, putatively material world of manifestation as being but "shadow appearances." And in the extremist Western view, in polar opposition to vivarta-vada, the realm of reality consists exclusively of noumena, as in the billiard ball ontology of Newtonian classical physics, where consciousness has no home.

In the past century, deepening and unifying scientific principles have been discovered that support broader, more informed theories than those attainable in Newton's time. These new bases for the sciences, in their prescriptions for the machinations of the physical world, including mental activity, elaborate upon the process posited centuries ago in the Shiva-Shakti Tattva. But science, by design, does not venture toward the crux of the Tattva, which elucidates the nature of the experiential product of Shiva's informational potentiality and Shakti's power to manifest it. In its highest expression, the macrocosm of Shiva-Shakti works together with the microcosms it infolds, so as to realize experiential vehicles of consciousness. According to the Shiva-Shakti Tattva, this process is the innermost essence of the universe.

A sutra says: "The Seer is pure consciousness but though pure, appears to see through the mind." This sutra stems from the fact that, from the top-down viewpoint of the Purusa, objects of consciousness *are* solely the result of intentionally practiced experiential control over, and anatomical transformation of, the body and mind. In this legitimate causal sense, the mind is part of Prakrti and is secondary to Purusa. Similarly, with respect to the Tattvic macrocosm, the essence of Shiva is taken to be pure consciousness, from which all of manifest reality springs forth by its inherent power of Shakti. From their top-down perspective, the ancients naturally analogized the fundamentality of Purusa consciousness in the microcosm, and the subservient role of physical Prakrti, with the ontology of the macrocosm.

The conception of the macrocosm as being fundamentally characterized by consciousness can be rationalized scientifically in terms of symmetry spaces in conjunction with the attractor of an Aristotelian "final cause" (the end toward which a thing naturally develops). The symmetry space of the macrocosm consists of potentialities, and since the *inevitable* product of cosmogenesis is the manifestation of Purusa consciousness in the symmetry space of a microcosm, from its first-person, intentionally causative perspective, consciousness *is* (teleologically) fundamental.

Some interpretations of the tattvic philosophy hold that Purusa is a passive, non-material witness to the activities played out in the mental regime of material Prakrti; again, in the words of Patanjali, "The Seer is pure consciousness but though pure, appears to see through the mind." In other interpretations, non-material Purusa is posited as a causal doer of actions. But what *is* a non-material entity? The search for answers to this question has a long history, and has been contemplated in all cultures since the beginnings of philosophy and religion. Plato, as noted, believed that all things possess an underlying *psyche* (Gr. soul) whose basis resides in the non-material world-soul, and any material appearances are illusory forms emanating from the causal (teleologic) source of transcendent reality. Other variants of this philosophy date back further, to the belief systems of China (Taoism), Japan (Shintoism), and India (Yoga; esp. the monistic forerunners of Advaita Vedanta). This class of panpsychic ontologies, which maintains the universal fundamentality of psyche, continued to develop in the West through the Greek, Neoplatonic, and Renaissance periods. Today, panpsychism is still considered to be a viable solution to the hard problem of consciousness,[101] which investigates the nature of the connection between the material and the experiential.

Another subject that has long been debated is whether Purusa-Prakrti and Shiva-Shakti are dualistic or monistic. In the dualistic Samkhya philosophy, experience and experienced are independently real, while in the monistic, panpsychic Vedanta, only experience is real, and the physical world is an illusion—*maya*. The issue of monistic panpsychism is related to the all-embracing symmetry of the macrocosmic universe. One could say that because the universe upholds consciousness as its inevitable expression of complexity, consciousness is the purpose, if not the essence, of reality. Further, the insight that complex and ultimately conscious systems emerge naturally from the fundamental holistic essence of the universe, has today gone beyond the natural sciences and experience-centric belief systems, into modern theology. The concept of "God" is increasingly understood to be the underlying source and essence of an ever-emerging universe:

> *Emergent monism is the ability of natural complexes to manifest new properties not associated with the components. New and distinctive kinds of realities at the higher levels of complex wholes have emerged that have causal power to make separated constituent parts behave in ways they would not do if they were not part of that particular complex system. The Being of God includes and penetrates all-that-is – so that every part of it exists within God and that God's being is more than, and is not exhausted by, it.*
>
> — Arthur Peacocke [102]

The evolutionary ascent of complex living beings has occurred along an extremum between order and disorder. On this anthropic pathway, self-organizing complexity has risen up continuously. The notion of an intrinsic animating source underlying the Dance of Shiva, or an implicate order of super quantum potential unleashing its active information upon the dynamic universe, captures that the source of complexity is imminent. This ontological principle appears in other areas of understanding as well. In Western theology for example, an ontology of God called theistic naturalism has been put forth, a belief which effectively adjoins science and the tattvic metaphysics of Brahman with respect to complexity and its emergence. In this theistic view, God is viewed as the perpetually imminent and creative principle that is not separate from the world, but is its very enfoldment in all of its aspects and reality.

Associating complexity and emergent monism with a concept of God brings science, tattvic philosophy, and theology onto a common platform, with a vital role for each: Science logically discovers and describes the properties of the physical, biological, and mental symmetry spaces. The tattvas incorporate another perspective, enhancing the same symmetry principles with the dimensionalities of conscious reality, the inevitable "apex of complexities in the world." Theology advocates the experience of, appreciation for, and personal relationship with the greater essence of the universe.

Sarvam khalv idam Brahman.
"Verily, all of this is Brahman."

Many cultures have deified the macrocosm, constructing relationships between the macrocosm and human microcosms. Invariably, the presumed macrocosmic god has been taken to be primary. This perspective is clearly stated in Genesis 1:27: "God created man in His own image." But, as indicated in the quote of Arthur Peacocke, where emergent monism is embraced, theologians are now also realizing the deep and unifying connection between a Shiva-Shakti type of macrocosm and emerging conscious microcosms—the connection between God and human beings.

> *The laws of nature would not be wholly arbitrary principles of interaction. They would be principles necessary to the fruition of a coherent, complex, organized and integrated universe. We could then speak of the supreme informational principle of the universe. In the basic physical foundations of the evolutionary process, there is a vector to the virtually inevitable development of conscious life.*
>
> *These considerations suggest the idea of a primordial consciousness that is ontologically prior to all physical realities, that contains the 'coded' information for constructing any possible universe. Whether or not one calls such a primordial consciousness God is partly a matter of taste, but if some notion of value is introduced, as a reason for actualizing some rather than other logically possible states, the notion of consciousness seems to be entailed.*
>
> — Keith Ward [103]

Above the scientific properties of consciousness that are discovered, and even beyond the knowledge gained and shared by the rishis of yore, is the personal wisdom that results from the practice of enlightened discrimination *(viveka),* the persistent reflection on the transcendent principle within Brahman that unifies us in all times and all places—the Shiva-Shakti Tattva. Based upon deep inward experience and developed intuition, the unity of Brahman, enfolding macrocosm and microcosm alike, was known by the seers. Their philosophy presaged the holistic ontology now attributed to both the quantum macrocosm and the living microcosm—confirming the ancient Stoic maxim, "As above, so below." Awareness of this truth bestows higher perspective, moral compass, and refuge, as we travel on the anthropic pathway. Within the universe of Brahman's symmetry, cycles of the macrocosm revolve through countless aeons of dissolution *(pralaya)* and activity *(sristi),* growing microcosmic flowers of consciousness along the way.

* * *

Part 2
Computers and Artificial Consciousness

10

Object Orientation

It has been more than 75 years since the deployment of the first programmable electronic computer at the University of Pennsylvania. The ENIAC (Electronic Numerical Integrator And Computer) filled an 1800 square foot room. Today's computers inhabit a chip of silicon no larger than a penny. While the size of computer hardware has undergone this dramatic collapse, processing power has exploded. A modern supercomputer boasts processing speeds a trillion times faster than the ENIAC.

Software, the logical procedures that harness the processing power of computer hardware, has also undergone a tremendous evolution since the early days of computer technology. A programming technique, called object orientation, has become indispensable to software design. Initially, software consisted of procedures (a sequence of instructions) specifying which hardware operations (register transfers, AND/OR gates, etc.) to perform on a set of data (in the registers, representing 0's and 1's). The procedures and the data were autonomous. On the ENIAC, procedures were executed in arrays of electronic vacuum tubes (off = 0, on = 1), while data were encoded on media like punched cards and paper tapes (unpunched = 0, punched = 1). Many of the great achievements of our computer age have been realized through the logic of *procedure orientation*.

In the earliest procedure-orientated computing systems, sets of coded data were small and manageable. But as the list of procedures lengthened into complex programs and the size of data fields grew, difficulties arose. System performance degraded, data was inaccessible to multiple users, interrelated program components could neither be easily modified nor reused, and development specialists became isolated. These limitations finally culminated in the "software crisis" of the 1960s. Software engineers responded by adding modular structure to both programs and data. This "structured revolution" led to four generations of software languages and to the birth of CASE (Computer-Assisted Software Engineering). On the data side of the equation, modular structures were introduced gradually through the evolution of hierarchical, network, and relational database management systems. These innovations provided many of the tools that would prove to be essential to the forthcoming development of object oriented systems.

Concepts of Object Theory

Object orientation does much more than remove inefficiencies associated with procedural programming. Object orientation allows a computer to simulate objects in the real world, complete with internal properties and behavior with respect to other objects. The first object oriented system was developed in Norway and aptly named Simula, which is an acronym for Simulation Language.

An object is a composite entity within which data are bound together, along with pointers to all of the procedures relevant to that data. The data define the particular characteristics or properties of an object, while the procedures represent its capabilities; i.e., the things the object "knows" how to do. This endogenous composition allows objects in the symmetry space of a computer (cyber-objects) to represent objects in many other types of symmetry spaces, including that of the physical world (world-objects), and to simulate their functional interrelationships.

The vocabulary used in object oriented computer science may be defined through the metaphor of a baseball game. In the following list of definitions, objects in the world of actual baseball players, or "world-objects," are represented by the symbol <w-o>; and objects in the generalized cyber symmetry space, or "cyber-objects," by <c-o>).

Properties

<w-o>: The attributes of an individual baseball player: batting average, running speed.
<c-o>: The data part of an object; .314, or 3 (seconds to first base), respectively.

Methods

<w-o>: The things a baseball player can do: batting, fielding, pitching.
<c-o>: The procedural component of an object.

Encapsulation

<w-o>: The combination of a baseball player's properties and methods within a baseball-player-object.
<c-o>: The packaging of properties and methods into a single object.

Class

<w-o>: All pitchers are members of the class "pitcher."
<c-o>: The set of objects of a particular type. Objects in a given class possess the same methods, but different values for their properties. An object is an instance of a class.

Inheritance

<w-o>: All classes of baseball players, such as pitchers and catchers, have certain methods in common, such as batting and fielding. Members of the class pitcher, for example, inherit the methods of batting and fielding from the class baseball player.

> Note: There may be other methods that are exclusive to pitchers. Methods such as throwing a curve ball apply only to those baseball players who are pitchers.

<c-o>: A relationship between two classes, in which one class, the "child," acquires methods from another class, the "parent."

Message <w-o>: A baseball player is only a potentiality until there is an interaction with other baseball players, an umpire, a fielder, and so forth. For example, when the umpire yells, "Play ball!" a message is issued that invokes the pitcher to pitch, the batter to bat, and the fielders to field. Through exchange of messages, world-objects communicate, inducing each other to activate their methods.
<c-o>: A signal from one object to another that instructs the receiving object to carry out one of its methods.

Polymorphism <w-o>: A given message can cause different baseball players to employ different methods. The message, "play ball" causes the pitcher to pitch and the catcher to catch.
<c-o>: The performance of different methods by different objects in response to a given message.

Anatomy of an Object

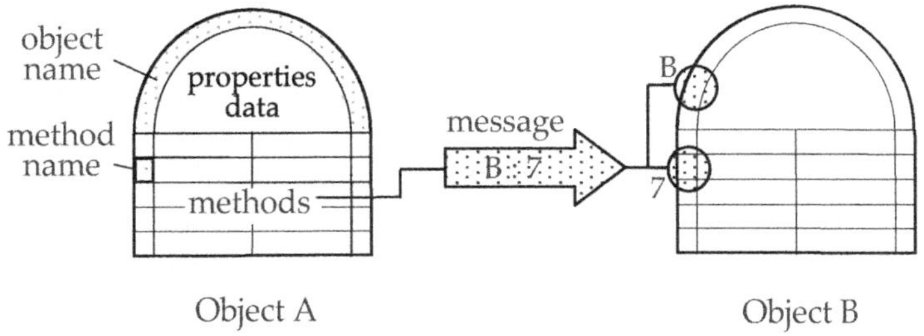

Figure 10-1 *Object Model*

A closer look at the anatomy of a cyber-object is depicted in Figure 10-1. In object A, a set of properties (data) is encapsulated with all methods (procedures) that may operate on those properties. In the figure, object A sends a message composed of an object name and a method name only.

If the object name, B, and the method name, 7, match those already encapsulated in a receiving object, the receiving object is triggered to invoke its method.

Object Food Chains

What makes object orientation so powerful, and what advantages do object based systems offer that elevate them above the traditional procedural computing algorithm? While many advantages derive from the fact that we can interface with cyber-objects as we do with the world-objects they simulate, additional benefits arise from the way objects work together.

Objects can be nested.
Objects can be linked.
Objects are horizontal and vertical.

The behavior of objects is reflective of their class membership. Objects belonging to the same class possess the same methods, or capabilities. Also, through inheritance, objects have access to the methods of ancestors further up the class hierarchy. When an object receives a message to invoke a method, it searches its own method library, as well as those of all its ancestors; when it is time to bat, a member of the "pitcher" class inherits the method of batting from its parent "baseball player" class. Another remarkable feature of objects is that, in addition to data, such as numbers and characters, as well as their self-defining methods, they can nest other objects. Objects containing other objects nested within themselves are composite objects, and their nesting can be implemented to any number of levels. The nesting hierarchy of composite objects differs from the class hierarchy of inheritance. In class relationships, the child is a sub-type of the parent; in contrast with nesting relationships, where objects exhibit very different behaviors, having no methods in common.

Objects can also be linked in such a way that they share property values. When a given object is modified, all other objects to which it is linked are automatically updated to reflect that change. Object linking and embedding protocols (nesting is called embedding in computer science) have been standardized to allow interoperability of third-party software products. Object interoperability further extends to the incorporation of procedural application code into object oriented systems through the artifice of "wrappers." Wrappers encapsulate the elements of procedural applications within an object shell. This capability has enabled businesses to smoothly migrate their procedure-based systems into object oriented environments. In the early days, the stratagem of wrappers furnished the business community, having a vital stake in the 100 billion lines of legacy procedural COBOL code which had accumulated by the 1990's, with a practical vehicle to enter the 21st century object oriented cyber-world.

All of these ways in which objects work together allow very sophisticated structures to be erected. Based on their complexity and the degree of functionality they provide, a hierarchy of categories has been demarcated for various levels of object grouping, including:

Class Libraries: Groups of individual objects (drivers: graphical user interface, print).

Components: Small groups of related objects that together perform basic functions (spell checker, drawing tool).
Applications: Groups of components that together produce powerful software tools (word processor, spreadsheet).

Suites: Groups of interrelated applications offered by major developers (Microsoft Office).

Chapter 10

Hardware Considerations

Successful implementation of software applications, whether procedure- or object oriented, is dependent on responsive hardware performance. Typically, the hardwired circuitries, and their basic electronic functions, are capable of supporting both procedure- and object oriented software. Program counters and registers, stack frame allocations, and instruction branching are common to both procedure- and method-based systems. Initially, insufficient processing power limited the performance of object oriented systems, as object functions require more operations per second than their procedural counterparts; message passing, for instance, is slower than conventional procedure calling.

The advent of microprocessor technology made the computational demands of object messaging practicable, even on inexpensive personal computers. In microprocessor technology, millions of transistors are etched into chips of silicon, or other semiconductive materials, in arrays of electronic circuitry. Logical patterns represented in the early PC arrays "flipped" at frequencies of hundreds of millions of operations per second. As increasing transistor densities and clock speeds continued to fuel performance gains, designers were able to implement object techniques to support proliferating high-end applications. Early steps forward in PC performance were made by Intel Corp. and Hewlett-Packard Co., who joined forces to design the P7 chip, which featured 64 bit wide data paths, massive operation parallelism, and Very Long Instruction Word (VLIW) computing—once again leapfrogging processor speeds and software capabilities. By 2000, home-user PCs performed with processor speeds of a billion operations per second.

The field of object orientation has sustained exponential growth since its inception, penetrating all quarters of computing and information technology. Objects have altered the substance and scope of software programming languages, database management systems, and end-user applications. But perhaps the most profound consequence of object oriented information technology has been the sociologic culture shift. The interaction between people and computers has changed. Initially, computers were utilized as machines dedicated to calculating answers to problems of mathematics and logic. Now, in the age of objects, computers enable people to interact in natural ways with simulated models of real world entities. The models respond to situations we impose on them, solving a wide variety of problems through use of their own knowledge sets and behavioral proficiencies.

Object Orientation in Nature

Object orientation is intrinsic to all interactions occurring in nature. At the fundamental level of physics, envariant quantum Darwinism engages the object orientation paradigm, where decohered particle-objects message with each other, as well as with the environmental-object (that encapsulates its own methods) in which they are nested. Throughout the ascent of classical material organization, object oriented processes have been instrumental in the production and maintenance of the fractal, intra-messaging object hierarchy that builds from decohered elementary particles, to atoms, to molecules, to autocatalytic sets, to life.

Object orientation is an essential feature in the abiotic realms of nature, as well as within and among living microcosms. In the intra-cellular symmetry space, flows of molecular RNA messengers instruct ribosome methods to biosynthesize the internal environment; and the replicating DNA object clearly represents object inheritance. In the human brain, nuclear objects in the ascending value system release neurotransmitter messages to distributed functional objects, instructing them to cooperate holistically; and the convergence-divergence nodes in the integration-differentiation hierarchy of the cortex implement polymorphic messaging.

Object orientation describes the process through which nature orchestrates the interactivity and evolution of complex living systems. Messaging within and among living, open-system microcosmic objects facilitates their transformational interdependencies, the discovery of selectively fit chemical and neural pathways, and genetic heritability—in the material regime. With respect to the processing, integration, and abstraction of the information essential to the evolution of consciousness, the necessary and sufficient biosemiotic platform is object orientated, with sign-messages being processed for their meanings relative to polymorphic interpretant-objects. The organic microcosm is the consummate embodiment of an object oriented system, operating in a symmetry space that brings forth the marvels of life.

* * *

11-1

Consciousness Transcendent

Our knowledge of the nature of reality is subject to limitations. However, the conscious mind can transcend them. To identify how the mind rises above, requires an assessment of what reality and knowledge are.

The Venn diagram below is a heuristic correlation of theories involving the nature of reality and our knowledge of it. The theories labeled 1, 2, and 3 are primary, representing the "world" (objective existence; macrocosm), the "mind" (subjective knowledge; microcosm), and mathematics, respectively. Many other theories which combine these basic notions have been developed over the course of history, and fall into the intersection sets labeled 4, 5, and 6. A brief definition of each is now provided, followed by a more extensive discussion.

Definitions

1- Realism: objective existence, independent of the mind.

2- Idealism: subjective existence, dependent on the mind.

3- Mathematics: patterns of nature and/or the mind.

4- Evolution of the mind.

- Epistemological Realism: observable characteristics exist in the observed object (macrocosm), independent of the observer.
- Epistemological Idealism: observable characteristics exist in the mind of the observer (microcosm), independent of the object.

5- Mathematical Realism: mathematical objects and truths are objectively real and independent of the mind.

6- Mathematical Idealism (Intuitionism): mathematical objects and truths are a creation of the mind.

Theories of Reality

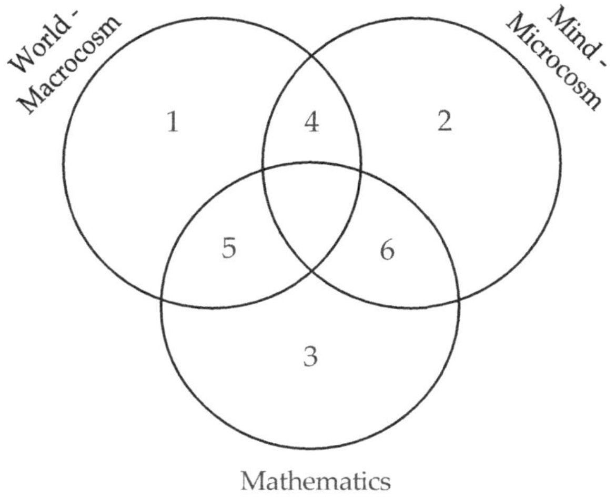

1- Realism	4- Evolution, Epistemology
2- Idealism	5- Mathematical Realism
3- Mathematics	6- Mathematical Idealism

Theories of Reality

1 - Realism (*objective realism, metaphysical realism*)

Reality is independent of the mind. Realists maintain that the external world has an objective existence independent of human observation and subjective experience. The mind may exist but it is not an essential attribute of reality. The ancient Greek philosophers, Plato and Aristotle, set the groundwork for a multitude of theories based on realism:

— *Platonic realism* is the view that "universals" (properties and principles held in common by individual things) are fundamental and more basic to the nature of reality than the (appearance of) things we actually perceive. Plato (427–348 BC), in his work *The Republic,* described reality as being abstract, perfect *ideal forms,* denying (in the *Republic's "allegory of the cave"*) the realism of the observed world of manifestations, which are imperfect "shadow appearances" of ideal forms; like observing images cast on the wall of a cave by a fire projecting the shadows of the real, but always hidden from view, ideal forms.

– *Aristotelian realism* (*Imminent realism*) is the view put forward by Aristotle (384–322BC) which, like Platonic realism, supports the realism of universals, but purports that their existence is dependent on the reality of the individual things which exemplify them.

2, 4 - *Idealism, Evolution, and Epistemology*

Idealism posits that reality is dependent on the mind. There are monistic philosophical theories dating back to the *Rig Veda* (2nd millennium BC), such as Vedanta, which claim that *only* consciousness is real (set 2 exclusively), and that objects of consciousness, physical and mental, are *maya* (Skt. illusion). It has been demonstrated in previous chapters that the mentally conscious mind is a product of *evolution*—having emerged in the interneuronal domain at the time of its holistic closure. Therefore, monistic idealism will not be considered here as an ontological basis of reality (but see Chapter 9, *Other Perspectives*).

The ancient Hindu Samkhaya philosophy is dualistic (intersection set 4), claiming that both the material world of nature (*Prakrti*, set 1) and consciousness (*Purusa*, set 2) are real. As mentioned above, we now understand that the mind is a product of evolution. We also understand from neuroscience and cognitive psychology that the process underlying consciousness is contingent upon an organism's knowledge-base of integrated information. Thus the intersection set 4 entails *epistemology*, the study of the nature, origin, and limits of knowledge, and, in the current context, the relationship between the mind and knowledge of reality.[ii]

---

ii The term "knowledge" refers to experienceable qualia that characterize integrated information produced by the brain ("The only source of knowledge is experience."—Albert Einstein.). To review: qualia are the source and essence of

There are two categories of epistemological theories (in set 4) regarding the relationship between knowledge and reality: epistemological realism and epistemological idealism.

— *Epistemological realism* claims that it is possible to obtain knowledge about mind-independent reality. It is the view that observable, therefore knowable, characteristics exist in the observed object, independent of the observer.

— *Epistemological idealism* is the idea that what one knows about an object exists only in one's mind. It is the view that observable, therefore knowable, characteristics exist in the mind of the observer, independent of the object. Note that epistemological realism and epistemological idealism presuppose the existence of both the world and the mind, but not that the world and/or the mind are fundamentally real.

---

imminent knowledge, and of the content of a dynamic knowledge-base. Qualia encompass a wide variety of informational types sufficient to fulfill the broad definition of epistemological knowledge, including: sensory-based perceptions (the redness of red, the taste of a strawberry), visceral feelings (pain), higher emotions (anger, love), semantic meanings (truth, mathematical concepts), imaginings, internal mental states obtained through introspection, self-reference (I versus it), memories, overviews of meaningful streams cognized through time (the understood thread of a conversation, the gist of a logical argument), and the feeling of intention (the resolve to pay attention).

Qualia represent varying degrees of the integration of bottom-up and top-down informational flows in the body and the mind of the microcosm. Bottom-up pathways have generally had sensory origins, having to do with the *effects* of objective, third-person interrelationships with the macrocosm, leading to the ascending channels of the nervous system. Top-down pathways have endogenous sources in the brain, having to do with the subjective, first-person vantage point of the mentally conscious organic microcosm. Top-down pathways may regulate, limit, and overcome the potential effects of bottom-up pathways, depending on microcosmic holistic exigency.

Definition of Reality

A working definition of reality based on our theory of the evolution and nature of consciousness, together with an evaluation of the various philosophical standpoints described above, may now be developed and stated. Such a definition of reality sets the context for theories that include mathematics.

In previous chapters, we have shown that symmetry is the underlying principle of reality, and that this symmetry is unstable (unstable tachyon supersymmetry at the big bang; Chapter 1). As a result of the instability, seven phases of evolution unfold through a cycle of cosmic evolution: 1) Symmetry, 2) Symmetry Breaking, 3) Complexity, 4) Holism, 5) Symmetry Integration, 6) Mental Consciousness, and 7) The Human Brain (and mind). Although the symmetry of the macrocosmic universe is broken, it is holistically unified (ontological interpretation of quantum mechanics; Chapter 17) by a set of ever-evolving interrelationships (where the term "interrelationships" encompasses all unified attributes of holism).

A primeval consequence of symmetry breaking is unidirectional time, along which complexity, microcosmic holism, and the mentally conscious mind *eventually* evolve. This implies that macrocosmic interrelationships are more basic to the nature of reality than one of its products, the microcosmic mind. Thus, the proficiency of the mind—as a complement of the brain—to perceive, acquire knowledge of, and specify the effects that macrocosmic interrelationships have upon it means that realism is more basic than idealism.

Reality is the holistic set of unspecified macrocosmic interrelationships that drives evolution toward complexity and the holistic microcosmic mind, which is uniquely capable of specifying effects of those interrelationships.

The *holistic set of unspecified macrocosmic interrelationships* is objectively real (realism) and not dependent of the existence of the mind. It is a property of the cyclic universe whose symmetry breaking, initiated at the big bang transition, underlies *evolution to complexity* and the emergence of the *holistic micro-cosmic mind*. The mind, in turn, *specifies effects of those interrelationships* (idealism). The capability of the mind to specify effects of macrocosmic interrelationships is based on the symmetry integrating cognitive processes and knowledge-base of the holistic microcosm. The modifier *effects* signifies that the specifications of the mind are about its experienceable qualia (mind-dependent epistemological idealism), and not directly about the properties and interrelationships of the universe (mind-independent epistemological realism) which exist on the other side of the horizon boundary that separates the microcosm from the macrocosm.

3 - Mathematics

The evolving set of macrocosmic interrelationships and/or the methods by which the mind perceives and specifies them, have mathematical properties. The correspondence between mathematics and the observed world has been recognized from antiquity. The Lebombo bone, discovered in Swaziland, is archeological evidence that humans have been counting things for at least 40,000 years. The fact that the Lebombo tally stick has 29 notches suggests a correlation with the tracking of lunar cycles. By 2600 BC, mathematical tables used for calculating quantities and properties of physical entities were etched into stone by the Sumerians. The Greek mathematician and philosopher Thales of Miletus (624 – 546 BC) first used mathematics to explain natural phenomena by means of the logic-based scientific method. And so forth, until today, when a new mathematical framework for M-Theory is being developed to unify quantum mechanics and general relativity.[104]

3, 5, 6 - Mathematical Realism and Idealism

Is mathematics a property of mind-independent reality that is *discovered* by the mind, or is it a *creation* of the mind that is superimposed on mind-independent reality, or is mathematics reality itself?

— *Mathematical realism* is the view that the truths and properties of mathematics are objectively real and independent of the mind (set 5). Pythagoras (570–495 BC) believed that the ultimate nature of reality is mathematics itself (set 3 exclusively), claiming "all things are number." Plato and Aristotle expounded theories of mathematical realism derived from Pythagoras. The philosophy of Platonism—Platonic realism applied to mathematics—asserts that abstract universal mathematical objects and principles exist in an objective reality, which may or may not be discovered through the agencies of the mind (the mathematical philosophy of *nominalism* denies the existence of mathematical objects and principles). Aristotle also framed a theory of mathematical realism, holding that properties such as symmetry and order are ontologically real. However, Aristotle's mathematical realism contrasts with Platonism in that the objects of mathematics, such as numbers, do not exist in an abstract world but can be physically realized.

— *Mathematical idealism (Intuitionism)* contends that mathematical truths and objects are a creation of the mind (set 6). Luitzen Brouwer (1881–1966), a founder of modern intuitionism says that the truth of a mathematical statement can only be conceived via a mental construction that proves it to be true.[105] Numbers, conceptual structures, and geometric configurations should be thought of in idealistic terms as constituting a sense-inaccessible realm of ideal objects.

Early 20th century discoveries, especially the uncertainty principle of quantum mechanics, led to deeper investigations and insights into the nature of reality, prompting a host of theories related to science and mathematics:

— *Positivism* is a philosophical school founded by Auguste Comte (1798–1857) which holds that meaningful knowledge can only be attained by scientifically verifiable logical reasoning based on sensory experience,

while knowledge derived from other sources such as introspection, intuition, and religious faith is considered meaningless.
—*Logical Positivism (Scientific Empiricism)* was a popular philosophy promoted in the Vienna Circle, an influential group of philosophers, logicians, mathematicians, and physicists who met regularly from 1924 to 1936 to discuss and debate ideas. Logical positivism proposes that all essential knowledge about a research issue can be gained through observation or logical proof. This philosophy ultimately failed because it unsuccessfully attempted to reduce the concept of clear and definite meaningfulness to the process of verification; the "verifiability criterion of meaning." This movement is mentioned here because Kurt Gödel (1906–1978), who quietly dissented from the idea, was a member of the Circle (which will be further elaborated upon in the section on Gödel's Incompleteness Theorem).

— *Gödelian Platonism (Conceptual Platonism)* postulates a hybrid of mathematical realism and intuitionism (sets 5 and 6) whereby Platonism is not primarily a kind of realism about mathematical objects existing independently of the mind, but rather a realism about the content of mathematical *concepts*, which implies Platonism about objects.[106]
"Gödel thinks that once we have grasped the concept "natural number" there is nothing further involved in the claim that natural numbers exist, because he thought the concept natural number itself has a real content, namely the existence of the natural numbers themselves."—Michael Potter [107]

— *Scientific realism* claims that the macrocosmic universe (set 1) to which scientific theories (set 4) refer is real, regardless of the correctness of the theories. Since the content of scientific theories includes scientific knowledge, scientific "realism" has an epistemological as well as an ontological dimension.
— *Mathematical empiricism* is a form of realism contending that mathematical facts are discovered by empirical research, just like facts discovered in the other sciences. Further, since mathematics is an essential, indeed

indispensable,[108] part of scientific theories, their realism necessarily encompasses mathematical realism.[109]

Mathematical Systems

Mathematical systems consist of two primary sectors: logical and conceptual. The logical sector, or *formal system syntax,* comprises meaningless symbols, strings of symbols called formulae, a particular set of strings called initial formulae, and mechanical rules for manipulating the initial formulae to generate additional formulae. The conceptual sector, or *deductive system semantics,* refers to a meaningful interpretation of the formal sector which underlies it. The deductive sector comprises meaningful symbols, axioms and inference rules (analogous to the initial formulae and mechanical rules of the formal sector, but with meanings attached), definitions of concepts and operations such as "even number," division," and "proof." When a formula is correctly generated, or proven, in the formal sector, the corresponding statement in the deductive sector is said to be "true."

More than 2,000 years ago, in his renowned work *Elements,* Euclid presented deductive systems for geometry and number theory based on the premise that *all provable statements are true.* For millennia, the converse was taken for granted: *all true statements are provable.* However, with his *Incompleteness Theorem* Gödel shattered the foundations of mathematical logic by demonstrating that *not all true statements are provable.* A mathematical system based on *deductive logic* (formal logic plus interpretation) such as number theory is therefore limited and *incomplete* in the sense that it can encompass a true statement, but not its proof nor refutation.

Gödel's Incompleteness Theorem joins with two other monumental discoveries of the 20th century that entail limitative results. These discoveries were in physics, where results are secured through the deductive logic of the scientific method: Chaos theory states that the future condition of complex dynamic systems, though governed by deterministic laws, cannot be predicted, and Heisenberg's uncertainty principle of quantum mechanics states that the position and velocity of a particle cannot be simultaneously known. Both of these limitative results arise from the impossibility to fully ascertain the initial condition of certain physical states: Chaotic systems are highly sensitive to initial conditions and quickly manifest in unpredictable patterns.[110] In the ontological interpretation of quantum mechanics, systems are associated with the past, reflecting the impossibility of knowing precisely all of the initial conditions necessary to calculate the present state; see Chapters 15, 17.

Within the field of mathematical logic, Gödel's Theorem was the first in a series of advances that uncovered fundamentally equivalent limitative results. Gregory Chaitin proved the incompleteness of mathematical systems by means of Algorithmic Information Theory.[111] In computability theory, Alan Turing solved the "halting problem" by proving that no computer program can determine, within itself, whether it will ever terminate.[112] And Alfred Tarski proved the inability of mathematics, and formal languages in general, to self-define the concept of truth. Tarski's work has profound implications for the philosophy of language.[113]

Each of these proofs was developed on the basis of deductive logic, long been accepted as a normative framework for capturing correct human reasoning. Deductive logic is characterized by *bottom-up* reasoning, beginning with basic assumptions and methodically building upon them to produce syntactic formulae and semantic statements. However, many psychological findings, such as the Wason Selection Task,[114] have demonstrated that deductive logic cannot serve as a descriptive language for the human inference process; there is at least some form of parallel processing in *cognitive logic*.[115] In particular, deductive logic breaks down in the context of self-referentiality. Gödel's proof, for example, constructs a self-referential statement called the Gödel sentence[iii] which effectively says, "I am unprovable"—akin to the Liar's Paradox, "I am lying," a statement that if assumed to be true *or* false leads to a contradiction.

How does deductive logic with its inherent limitative results regarding self-referentiality relate to the human brain and mind? [116] Firstly, human beings are holistic *self-referential* microcosmic systems, and the brain is an organ in the service of that holism. Secondly, the brain and mind operate on the basis of cognitive logic and integrated information. Integrated information is produced via a nodal network of profusely interconnected bottom-up *and* top-down, i.e., *cyclic,* neuronal pathways, according to the parameters of cognitive logic which is characterized by self-referential, holistically causal, "axioms and inference rules." Thirdly, through the process of neuronal group selection, certain representations of integrated information are experienceable by the dedicated ultrasensory collective neuronal structure that possesses the emergent property of consciousness.

Deductive logic is a product of cognitive logic and thus a creation of the mind. Being the progenitor of deductive logic, cognitive logic and consciousness stand above the range of deductive logic and therefore transcend it, as well as our knowledge of reality based on deductive logic. However, the same concepts conceived by the conscious mind for implementing a system of deductive logic can be applied to the deductive system itself, in which case the deductive system is treated as the object to be reasoned about. For mathematics—our descriptor of reality based on deductive logic—this kind of reasoning is termed *metamathematics;*

---

iii The Gödel sentence can be represented mathematically by the self-referential statement U ("U"), where U is a complicated expression developed by Gödel that effectively means 'unprovable.' U ("U") therefore states that the "unprovable" is unprovable. Gödel then proved that U ("U") itself is neither provable nor refutable; i.e., that the Gödel sentence is undecidable.

that is, metamathematics is the study of mathematics itself using deductive logic.

Through an extraordinary feat of ingenuity—a transcendent reasoning power of cognitive logic and the conscious mind—Kurt Gödel was able to prove metamathematically that mathematical systems based on deductive logic are irreconcilably limited and incomplete. Gödel proved in metamathematics that while mathematics is a *proof system* that can construct statements which are true, it *cannot prove* that they are true. We now present a graphical proof sketch of Gödel's Incompleteness Theorem.

* * *

11-2

Gödel's Incompleteness Theorem

In 1931, a young Austrian genius named Kurt Gödel invalidated a belief that had been held since the time of Euclid more than 2000 years ago. Euclid instituted a method for proving the truth of mathematical statements. In Book I of his masterpiece, *Elements,* Euclid proved the truth of the Pythagorean Theorem, and in Book IX—which involves number theory, the branch of mathematics germane to Gödel's work—Euclid proved the truth of the statement that there are an infinite number of primes. The central idea underpinning Euclid's method is that *all provable statements are true,* and for millennia the converse of this tenet was taken for granted: *all true statements are provable.* However, with his *Incompleteness Theorem,* Gödel shattered the foundations of mathematical logic by demonstrating that *not all true statements are provable.*

Euclid's method is familiar today. High school geometry textbooks are essentially a replication of Book I. We recall his "axiomatic method" from geometry class: assign names to intuitively understandable concepts such as "point" and "line" (*primitive notions*), assume certain basic statements about primitive notions to be true because they appear to be self-evident (*axioms*), and manipulate those axioms with truth-preserving logical operations (*inference rules*) to produce, or "prove," more true statements (*theorems*). Once a theorem has been proven, it joins with the axioms as being a true statement available for further deductive reasoning and the development of a rich mathematical system.

The evolution of logic and mathematics has an interesting history dating back to Aristotle (384-322 BC), the "Father of Logic." In *Prior Analytics,* Aristotle employed symbolic variables to represent logical arguments called syllogisms. Syllogisms are based on premises assumed to be true.

For example, the syllogistic argument: all humans are mortal, Socrates is human, therefore Socrates is mortal, would be represented symbolically by: all H are M, S is H, therefore S is M. In a logical argument, the symbols and rules for manipulating them are termed *syntax*, and the meanings assigned to them, such as truth or mortality, are *semantics*. Aristotelian logic is always a mixture of syntactic and semantic notions.

Aristotle's student, Theophrastus (371–287 BC), described *modus ponens*, which is now (but not then) understood to be a semantics-free principle of pure logic. *Modus ponens* is symbolized by: A→B, A, ∴ B. This argument says: If A then B, A is the case, therefore B is the case. A meaningful interpretation was generally assigned to this syntactic symbolism. For example: If it is raining then it is sunny, it is raining, therefore it is sunny. Note that *modus ponens*, in contrast to a syllogism, is a *valid* argument form regardless of whether the premises A and B are actually *true* (which they are not in this example). And, it was during this prolific period that Euclid (325-265 BC) presented his axiomatic systems of mathematics solely in the common Greek vernacular of the day.

The conflation of syntax and semantics continued until the 1800s when it became evident that both geometry and number theory required stronger foundations. In geometry, emerging ideas about curved space—which would later serve as the basis for Einstein's theories of relativity—needed to be developed; and in number theory, subtle paradoxes were being uncovered. It was realized that the informal languages used to articulate deductive mathematical systems were imprecise, and a movement to ground mathematics in logic became a central issue, engaging an intense group effort. In this endeavor, Gottlob Frege advanced a modern system of mathematical logic; Giuseppe Peano formulated a list of eight axioms, three of which are now considered to be in the realm of syntactic logic, while the other five refer specifically to number theory; and Bertrand Russell and Alfred North Whitehead wrote *Principia Mathematica*, a three volume tome based largely on Frege's logic and Peano's symbolism. But these attempts (and others) invariably entangled syntax and semantics, falling short of the goal to set the foundation of mathematics in pure logic.

The Vienna Circle

Frege, Peano, Russell, Gödel, David Hilbert, and Albert Einstein were members of the "Vienna Circle," an influential group of philosophers, logicians, mathematicians, and physicists that convened periodically to discuss and debate ideas. These meetings were supplemented by other international assemblies and frequent interpersonal correspondences. Noteworthy communications transpired between: Russell and Peano—resulting in Russell's adoption of much of Peano's symbolism in Principia Mathematica; Frege and Russell—from which Russell adopted much of Frege's logic, but whereby Frege described himself as having been devastatingly "surprised beyond words" and "thunderstruck" that his system of logic based on set theory was toppled by "Russell's Paradox" (Does the set of all those sets that do not contain themselves contain itself?); Hilbert and Einstein—leading to the still ongoing debate over who deserves priority for relativity theory; and Frege and Hilbert—on the separation of syntax and semantics vs. a "universalist" conception of logic. (In later years, Gödel and Einstein, both having fled Nazi Germany, would reunite and become close friends at Princeton University's Institute for Advanced Study where they took daily walks across the campus together, about which Einstein commented that he went to work "just to have the privilege of walking home with Kurt Gödel.")

In 1899, a strictly *formal* treatment of Euclidean geometry (based exclusively on syntactic symbolism and procedures, devoid of semantic connotations) was achieved by David Hilbert in his book *The Foundations of Geometry*. Hilbert also prepared a list of 23 unsolved problems as a challenge to future generations of mathematicians (10 of which have now been solved). The second problem, referred to as "Hilbert's Program," asked whether *all* of mathematics *could*, in principle, be based completely upon a formal system of logic, or is mathematics inherently *incomplete;* that is, are there any statements, or *propositions,* that can be properly expressed in a formal system but which are *undecidable* (neither provable nor disprovable within that system)? In his 1931 paper, *On Formally Undecidable Propositions of Principia Mathematica and Related Systems 1,*[iv] Gödel provided the stunning answer: Yes! And most ingeniously, he attained this result by proving that the proposition, "the unprovable is unprovable," is undecidable.

Gödel's Proof

Gödel's paper specifically addresses number theory, the field of mathematics dedicated to the study of the natural numbers: 0, 1, 2, 3... However, while his paper concerns only the natural numbers *per se,* its consequences are far-reaching. Advanced realms of mathematics, including the real numbers (negatives, fractions, π, etc.), imaginary numbers, algebra, calculus, and analytic geometry (the representation of geometry in algebraic terms) are all products of number theory.[117] Furthermore, these higher realms of mathematics, built from the natural numbers, describe the laws of nature. Consequently, in developing the formal system upon which he ascertained his results, Gödel ensured a

---

iv Gödel titled his paper *Über formal unentscheidbare Sätze der Principia Mathematica und verwandter Systeme 1.* In reference to his paper, the German word *Sätze* is generally translated to English as *Propositions.* But, to be precise, a logical proposition is defined in English as a statement which has a truth value, i.e., it is either true or false. Gödel's proof directly concerns formal undecidability and is syntactic, independent of semantic truth values (although the proof has important semantic implications, such as *not all true statements are provable*). We will use the comprehensive word "statement" in our discussions except where the word "proposition" is contextually pertinent.
The number "1" at the end of the title refers to Gödel's intention to publish a clarifying extension to his paper. His proof was so widely accepted, however, that the effort became unnecessary, and part 2 was never forthcoming.

reciprocal correspondence with the properties and relations of the natural numbers.

Gödel's proof of the Incompleteness Theorem was innovative. He constructed a self-referential statement in the language of mathematics, called the *Gödel sentence*, which effectively says, "I am unprovable." He then showed that the Gödel sentence is *unprovable;* and therefore what the Gödel sentence explicitly asserts—its own unprovability—is *true*. It is in this sense that a mathematical system which can encompass a true statement but not its proof (nor its refutation) is said to be *incomplete*. A few years after Gödel presented his Incompleteness Theorem, the Polish logician Alfred Tarski advanced the conceptual breakthrough by showing that within any formal system, the very notion of *truth is undefinable,* further exposing the intrinsic limitations not only of mathematics but of language in general.

* * *

We now present a graphical proof sketch of Gödel's First[v] Incompleteness Theorem.[118] The proof sketch contains nine figures which represent the primary means of description. To consolidate the various text styles, abbreviations, and symbology used in the figures, the following Legend is also provided:

---

v Gödel proved the following three important theorems: The *Completeness Theorem* (concerning individual propositions in any formal system) says that any proposition built up from the axioms is provable. The *First Incompleteness Theorem* (concerning a formal system itself) says that there are propositions which are expressible but not provable (nor disprovable) from the axioms of that system). The *Second Incompleteness Theorem* (also concerning a formal system) says that it cannot prove its own consistency; i.e., no proposition is both provable *and* disprovable).

Chapter 11-2

LEGEND

TEXT

PURPOSE OF TEXT	*TEXT STYLE — EXAMPLE*
HEADINGS	SMALL CAPS— P – FORMAL SYSTEM - SYNTAX
Descriptive information	lower case normal— initial formulae, mechanical rules
Arithmetical expressions in NT	lower case normal— $\text{Div}(x, y) =_{def} \exists z\, [(z \leq x) \wedge (x = y.\, z)]$
Gödel numbers; Quine quotes	lower case normal— $\ulcorner \varphi (x) \urcorner$
Auxiliary commentary	*Italics—* *Develop NT statement b that means B*
Typographical strings in P	**Upper case Bold—** **X, Y, PRF (x, y)**
<u>Numerals in P</u>	**lower case bold, <u>underscore</u>—** **x, y, <u>⌜ φ (x) ⌝</u>**

ABBREVIATIONS

M – Mathematical system
MM - Metamathematical system
P – Gödel's formal system
NT – Number theory deductive system
GN – Gödel Number coding
PR – Primitive recursive
PM – Principia Mathematica

LEGEND

SYMBOLS

SYMBOL	*DESIGNATES*
$\vdash$	is provable (short for P $\vdash$)
$\nvdash$	is unprovable
$\therefore$	therefore
=def, $\equiv$	definition
Φ (x)	formula
$\wedge$	and
$\vee$	or
$\neg$	not
$\ulcorner \dots \urcorner$	Quine quotes
S	successor function
ω	omega (consistency)
$=$	equals
$\neq$	does not equal
$\leq$	less than or equal to
$\rightarrow$	implies
$\leftrightarrow$	if and only if
$\forall$	for all
$\exists$	there exists
$\nexists$	there does not exist
$\otimes$	multiplication
$\in$	is an element of
$\mathbb{N}$	natural numbers

Within the figures, typographical strings in P, such as **PRF (x, y)**, are written in bold text font, as specified above. But in the narrative outside of the figures, where the discussion is conceptual and informal, light text font is used, e.g., Prf (x, y).

FIGURE 1 - MATHEMATICS AND METAMATHEMATICS

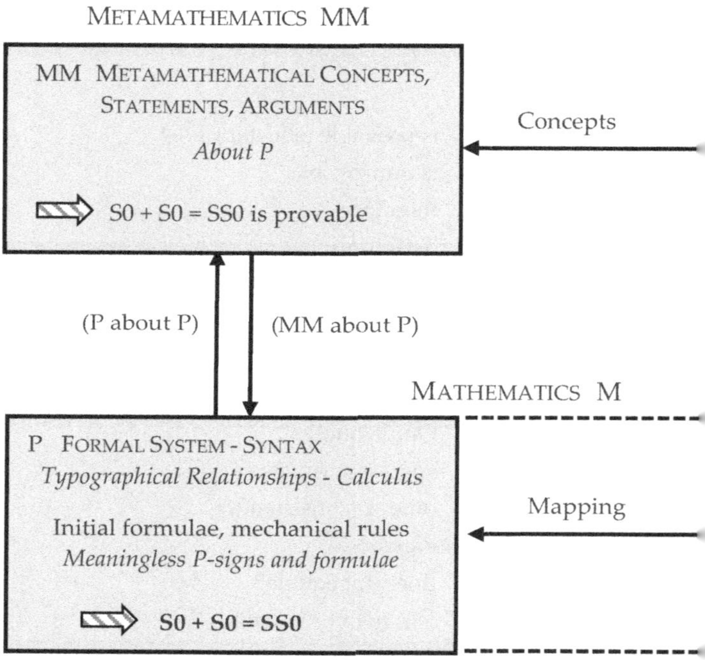

Description of Figure 1

Figure 1 illustrates the relationship between mathematics and metamathematics.[119] It depicts the underlying foundation upon which Gödel developed his proof. The shaded boxes in the diagram demarcate the four main sectors of the mathematics-metamathematics relationship; namely, the three sectors of mathematics M which are: P (Gödel's formal system of logic), NT-Semantics, and NT-Operations (where NT refers to the 'deductive system' of number theory [single quotation marks reference text as it is explicitly written in the figures]), and metamathematics MM which is the study of mathematics itself. The labeled arrows connecting

FIGURE 1 - MATHEMATICS AND METAMATHEMATICS

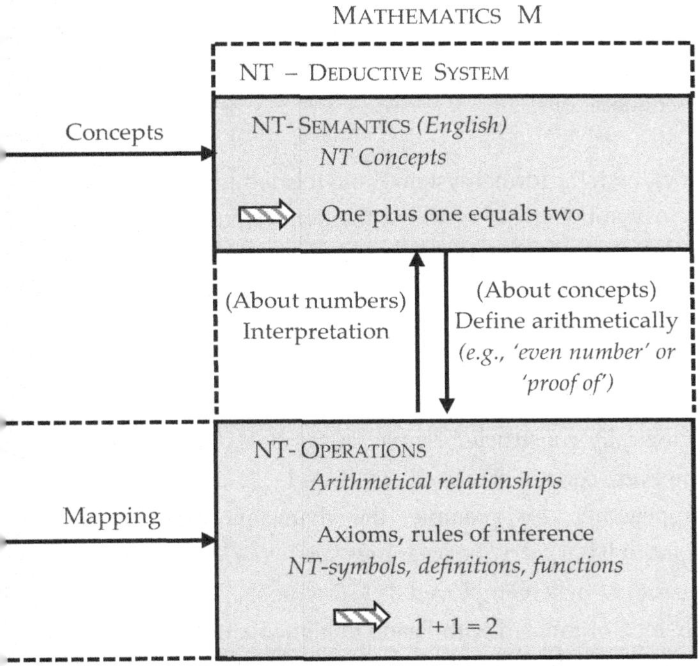

the boxes specify the kinds of relationships which associate the corresponding sectors. Within each of the sectors, a wide hatched arrow points to a typical example of how the subject of 'one plus one equals two' is expressed.

—As shown in the boxes for P and NT-Operations, the foundational principals and rules of the formal system P are termed 'Initial formulae, mechanical rules,' while for NT-Operations they are termed 'Axioms, rules of inference.' The key distinction is that initial formulae and the mechanical rules (Gödel's term) for manipulating them are meaningless, while axioms and rules of inference connote meanings. However, the terms axioms and rules of inference are so widely used (including in Gödel's paper) with reference to both P and NT that we *will* adopt this context-dependent convention, keeping in mind that only in NT do they carry semantic connotations.

—In mathematics, both the formal system P and the deductive system NT employ the use of symbology. We will use the term "signs" to designate the typographical characters of P, while the term "symbols" will be reserved for the characters of NT (despite traditional and current usage, such as the term "symbolic logic" to describe P-syntax). Again, the key distinction is that the signs of P, including logically constructed strings of signs called 'formulae,' are meaningless; whereas the symbols of NT, including logically constructed strings of symbols, have meanings attributed. However, certain characters in P and NT are, by design, identical in appearance; for example, the characters ∨ ¬ ∃ S are meaningless signs in P but carry the meanings "or," "not," "there exists," and "the successor of," when employed as NT symbols. Many symbols, or variations thereof, originated with Peano in his semantic treatment of number theory, and were then coopted by Russell into PM in his attempt to formalize mathematics, and are now used to represent signs in strictly formalized syntactic systems, such as P.

—The character S shown in the box for P warrants special mention; it is called the *successor function*. Along with 0 (zero) and n (number), S is one of the three primitive notions postulated by Peano as the basis for NT. The meaningful quantity five, for instance, may be represented in NT by the symbol SSSSS0 (the fifth successor of 0), but is usually represented by the symbol 5 (in decimal notation); these symbols are termed *numbers*. The successor function S is also employed in P to generate meaningless strings of signs in accordance with the mechanical rules of P; these signs are termed *numerals*. Note that only the two signs S and 0 are required to specify any numeral in P, regardless of how large its quantitative interpretation in NT.[vi]

Together, P and NT constitute an *interpreted formal system* for mathematics. But, in general, a given syntax may carry no interpretation at all, or it may have multiple interpretations whereby each interpretation pertains to some particular, meaningful subject matter, such as NT. The nature of the meaningful semantic 'concepts' that supervene upon the strings and operations of NT is described in the boxes for NT-Semantics

---

vi Peano showed that the entire theory of the natural numbers NT could be derived from three primitive notions, four axioms of pure logic, and five axioms dedicated to the NT interpretation.
The three primitive notions are: 0, number, and successor.
The four axioms of pure logic are:
1- For every individual x, $x = x$. That is, equality is reflexive. (Under the interpretation of NT, individuals are the natural numbers, so that x, y, z may be denoted by n where $n \in \mathbb{N}$ as shown in the text example.)
2- For all individuals x and y, if $x = y$, then $y = x$. That is, equality is symmetric.
3- For all individuals x, y and z, if $x = y$ and $y = z$, then $x = z$. That is, equality is transitive.
4- For all x and y, if y is an individual and $x = y$, then x is also an individual. That is, the individuals are closed under equality.
The five axioms specific to NT are:
1- 0 is a number.
The remaining axioms *define* the arithmetical properties of the natural numbers.
2- Every natural number x has a successor denoted by Sx. We can construct the sequence of terms `0', `S0', `SS0', `SSS0'. . .(called "numerals" in the formal system P, which correspond to the "numbers" 0, 1, 2, 3, 4 . . . in the NT interpretation).
3- No two numbers have the same successor; if $x \neq y$, then $Sx \neq Sy$.
4- 0 is not the successor of any number; $(0 \neq Sx)$.
5- Any property that belongs to 0 and also to the successor of any number, belongs to all numbers (induction); $\phi(0) \wedge \forall x[\phi(x) \rightarrow \phi(Sx)] \rightarrow \forall x\phi(x)$, where ϕ symbolizes "property," $\wedge$ symbolizes "and," $\forall$ symbolizes "for all," $\rightarrow$ symbolizes "implies."

and NT-Operations, and by the labeled arrows linking them. In particular, the idea of the truth, falsehood, or undecidability of a statement, being meaningful concepts, belongs to the semantic domain; but a syntactic composition of signs in a formal system that is true in one interpretation may be false or undecidable in another, depending on the meanings assigned to the symbols of the interpretation.

Normally, the syntax of a formal system is designed with an intended semantic interpretation in mind.[120] Gödel says of the syntactic system P that he devised for his paper, "P is essentially the system obtained by superimposing on the Peano axioms the logic of PM."[121] The Peano axioms refer to the deductive system developed by Peano to serve as the foundation for number theory, which supports a semantic domain.
For example, $n \in \mathbb{N} \rightarrow Sn \in \mathbb{N}$ is a Peano axiom
The intended interpretation of this symbolic statement is: if n is an element of the set of natural numbers $\mathbb{N}$, then the successor of n is an element of the set of natural numbers $\mathbb{N}$. Here, the symbology is specifically designed to accommodate number theory.
In Gödel's quote, PM refers to the formal logic (part) of *Principia Mathematica* that presides over its number theoretic interpretation.
For example, $(p \vee q) \rightarrow (q \vee p)$ is an initial formula of PM
If either p or q then either q or p
But doesn't this English language statement, "If either p or q then either q or p" assign a meaningful interpretation to the syntax of the formal (part of) the system PM, which is supposed to consist only of meaningless uninterpreted signs? The syntactic signs in this example are naturally intended to express the meaningful idea, "If either p or q then either q or p." However, this idea is *abstract* and does not entail a specific, *concrete* interpretation, such as a number theoretic statement. In a formal syntactic system, such as P or (the logic part of) PM, terms like p and q are to be understood as abstract placeholders for unspecified semantic statements, and together with the properties and relationships (e.g., $\vee$, $\rightarrow$) among

these placeholders, they constitute logical principles which are devoid of meaningful content.

The formal syntax and arithmetical operations used by Gödel are tailored for his proof and are more concise than those of PM and Peano. Indeed, Gödel's quote goes on to say, "The addition of the Peano axioms...serves only to simplify the proof and can in principle be dispensed with." (because they can be derived from PM). For efficiency, Gödel employs only three of Peano's axioms[122] directly, but adds eight "axiom-schemata" (which, like placeholders, apply to an unlimited number of cases). The advantage of his axiom system is that only two rules of inference are required for his proof, namely, the conditions for a formula to be an "immediate consequence" of either one or two formulae previously established by the axiomatic method.

The basic 'axioms, rules of inference' shown in the box for NT-Operations (with support from NT-Semantics) are enhanced by the addition of meaningful symbols, definitions,[vii] relations (among numbers), and func-

---

vii Definitions by recursion are constructed sequentially by working from the bottom up: addition is defined in terms of the successor function, multiplication in terms of successor and addition, factorial in terms of successor and multiplication, and exponentiation, also in terms of successor and multiplication.

Addition, multiplication, and a hierarchy of functions are defined through the process of recursion as follows:

Addition is a function that maps two natural numbers to another one. It is defined as:, $x + 0 = x$ or $x + Sy = S(x + y)$ (two equations, for $n = 0$ and $n \neq 0$)

Given addition, multiplication is a function defined recursively as:

$$x \otimes 0 = 0 \text{ or } x \otimes Sy = (x \otimes y) + x$$

More recursively defined functions can then be developed from addition and multiplication, e.g.:

Factorial, $0! = 1$ or $(Sy)! = y! \otimes Sy$

Exponentiation, $x^0 = S0$ or $x^{Sy} = (x^y \otimes x)$

tions (operations that map numbers to other numbers), all of which can be represented by P. Examples: the definitions of the concepts "even," "prime," and "proof," the relationship between x and y in the expression x^y, and the functions (and symbols) for addition (+) and multiplication (⊗), all increase the richness and strength of the system. In particular, the arithmetic functions of addition and multiplication are essential to Gödel's proof, as both are required to confer to his system the *sufficient strength* (complexity) to be incomplete, whereas, for example, Presberger arithmetic[123] that includes addition but not multiplication is not incomplete because all of its propositions are decidable. There are actually four mutually constraining properties of a mathematical system, so that Gödel's theorem must be qualified as follows: no system that is axiomatizable, sufficiently strong, consistent (cannot both prove *and* disprove a proposition), and decidable[124] can be complete. Thus, in demonstrating by Gödel's First Incompleteness Theorem that a system is incomplete because it harbors undecidable propositions, we must stipulate *a priori* that it is axiomatizable, sufficiently strong, and consistent. Conversely, if a system is axiomatizable, sufficiently strong, and decidable, then it cannot prove its own consistency—which is Gödel's Second Incompleteness Theorem.

Description of Figure 2

Metamathematics MM is the study of mathematics. More precisely, metamathematics is *about* the formal sector of mathematics P. Figure 2 illustrates the procedure for converting a metamathematical statement B in English (about a P-formula A) into an equivalent metamathematical statement B in P-syntax. Representation of the statement B in P-syntax is imperative, for then it is grounded in pure logic and not subject to vagueness, ambiguity, or the idiosyncrasies of an interpretation. This procedure lays the groundwork for achieving a crucial step in Gödel's proof: developing a particular P-formula B (designated by G, the Gödel sentence) that says there is not a proof of another particular P-formula A (designated by U).

The procedure for converting a metamathematical statement from English, or any informal language, to P is circuitous, as portrayed in Figure 2 by the cycle of pathways labeled by the circled numbers 1 through 5. Pathway 1 depicts an MM statement B (in English) about a P-formula A; while Pathway 5 depicts the result of the conversion procedure which is the MM statement B (in P-syntax) about the P-formula A. Pathways 2-4 specify two of the key tactics employed by Gödel to implement the procedure, which are Gödel Numbering and Primitive Recursiveness.

Gödel Numbering:
Gödel numbering is a mapping between P and a subset of the prime numbers in NT.[viii, 125] Signs, formulae, and sequences of formulae in P can

---

viii *Gödel Numbering*: Each of the primitive signs in Gödel's system P: 0, S, ¬, ∨, ∀, (,) are assigned the NT natural number: 1, 3, 5, 7, 9, 11, 13, respectively, and every variable: x, y, z is assigned a prime number greater than 13 with an exponent greater than zero which indicates its type (sign, formula, sequence of formulae). Hence, a formula, which is a finite sequence of $n \geq 1$ signs, is also a sequence of natural numbers denoted by $m_1 \ldots m_n$, where each m_i is the natural number that was assigned to the ith sign in the formula. The Gödel number of a formula is then defined as $p_1^{m1}x \ldots xp_n^{mn}$, where each p_i denotes the ith prime number. This number is unique due to the Fundamental Theorem of Arithmetic, which states that every natural number has a unique decomposition into its prime components. A finite sequence of $n \geq 1$ formulae is also a sequence of Gödel numbers denoted by $g_1 \ldots g_n$, where each g_i is the Gödel number of the ith formula in the sequence. As a result, the Gödel number of a finite sequence of formulae is $p_1^{g1}x \ldots xp_n^{gn}$.

FIGURE 2 - METAMATHEMATICAL SYSTEM, FORMALIZATION PROCEDURE

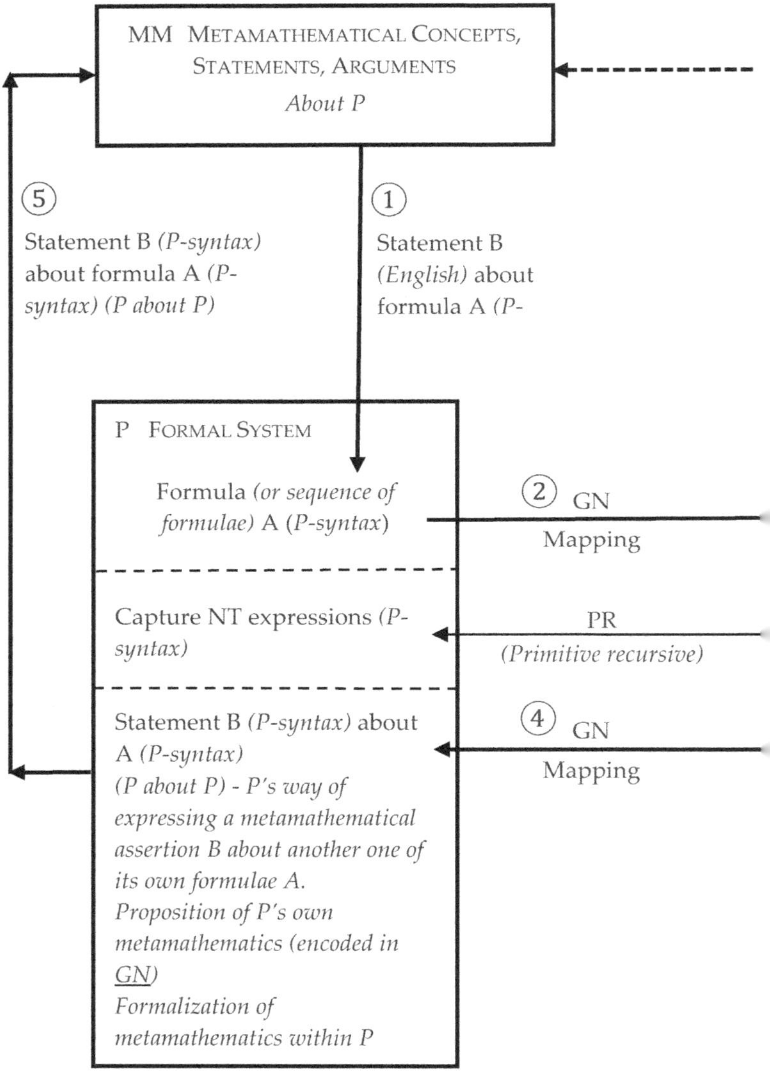

FIGURE 2 - METAMATHEMATICAL SYSTEM, FORMALIZATION PROCEDURE

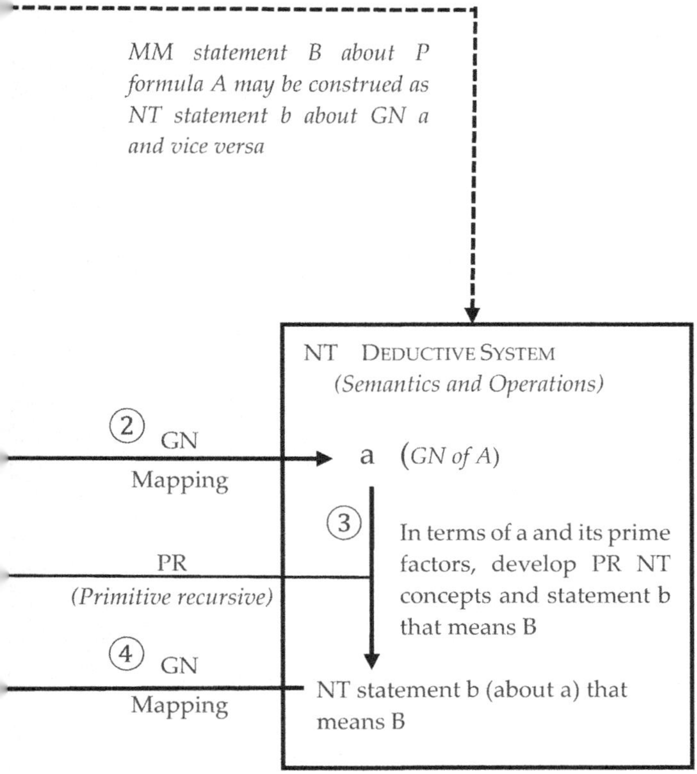

be coded uniquely into these "Gödel Numbers" GN in NT. Conversely, Gödel numbers in NT can be decoded into P syntax. The ultimate goal of Gödel numbering is to 1) encode a P-formula (or sequence of P-formulae)into a GN in NT, 2) process that GN and its prime factors within NT so as to develop a meaningful statement in NT, and 3) convert the new NT statement into a new P-formula. To someone who knows the Gödel numbering code, the new P formula can be deciphered to obtain the meaning about the original P-formula imbedded within it.[126] Thus, from a metamathematical perspective, the new P-formula "says" something about the original P-formula.

This procedure is detailed in Figure 2 as follows:
—Pathways 1, 2; a sign, formula, or sequence of formulae A in P, which is the object of an MM statement B in English, is encoded into the GN *a* in NT (*a* and *b* are in italics in this section to avoid ambiguity).
—Pathway 3; the GN *a* and its prime factors are then operated upon in NT to develop a concept-bearing statement *b* in NT symbology that means B.
—Pathways 4, 5; the NT statement *b* is then transported directly into P, producing an MM statement B about A in P-syntax.[127] The NT symbolism used to express the statement *b* can be transported into P, including definitions and abbreviations developed in NT, because they are primitive recursive.[128]

Primitive Recursiveness:
—Pathway 3; it is essential that each step in the development of statement *b* is primitive recursive PR. A PR statement is one that is built up from the axioms by the rules of inference of NT such that a correspondence with the initial formulae and mechanical rules of P is assiduously maintained, thereby grounding each NT step in the pure logic of P.[129]

Description of Figure 3

Figure 3 shows how to construct a simple metamathematical statement in the formal system P. In this example, attention is paid to the conceptual aspects of the procedure. The main sectors of the MM-M relationship have been partitioned into the two dashed boxes labeled *Concepts* and *Conceptual Objects*, in which the term "concepts" refers to meaningful understandings of the mind and "conceptual objects" to the meaningless typographical characters and systematics upon which the concepts supervene. Specifically, the metamathematical statement in the MM-box which is to be formalized in P-syntax is "The first sign in the P-formula S0 + S0 = SS0 is S." This is a meaningful statement in English (the concept: "the first sign… is…") about a meaningless string of signs already in P-syntax (the conceptual object: S0 + S0 = SS0). The labeled arrows joining the partitions describe their interrelationships.

The formalization procedure, which concludes in a P-formula B about another P-formula A, unfolds along a cycle of seven pathways as follows:
—Pathway 1; the statement B to be constructed in P-syntax is identified in the MM box: "The first sign in the P-formula S0 + S0 = SS0 is S." [*This is statement B in MM.*] This statement involves an MM concept (the position of a sign in a string) about a P-formula A (S0 + S0 = SS0) which is its conceptual object.
—Pathway 2; formula A is calculated to have the GN *a* with respect to NT. Decompose *a* to identify its first prime factor 2^7 which corresponds to the first sign S in P-Formula A.
—Pathways 3, 4 (NT semantic concept sub-loop); exercising intuition and ingenuity (similar to solving a word problem in algebra), find an expression in NT-Semantics stating that the first prime factor in *a* codes the first sign S in A: 2^7 is a factor of *a*, and 2^8 is not a factor of *a*. [*This is statement B in NT-Semantics.*]
—Pathway 5; again exercising intuition and ingenuity, develop an NT statement *b* in NT-Operations that (symbolically) means B: $\exists z\ (a = z \otimes 2^7) \wedge \neg \exists z\ (a = z \otimes 2^8)$ [*This is statement B in NT-operations.*]
—Pathways 6, 7; the NT statement *b* is then converted into P syntax, producing an MM statement about A: ∃Z (SSS …SSS0 = Z ⊗ SSS … SSS0) ∧ ¬ ∃Z (SSS …SSS0 = Z ⊗ (SSS … SSS0), where the quantity of S's in each of the four numerals represented by SSS …SSS0 in P equals the corresponding number in NT-Operations. [*This is statement B in P.*]

FIGURE 3 - METAMATHEMATICAL SYSTEM, CONCEPT LOOP

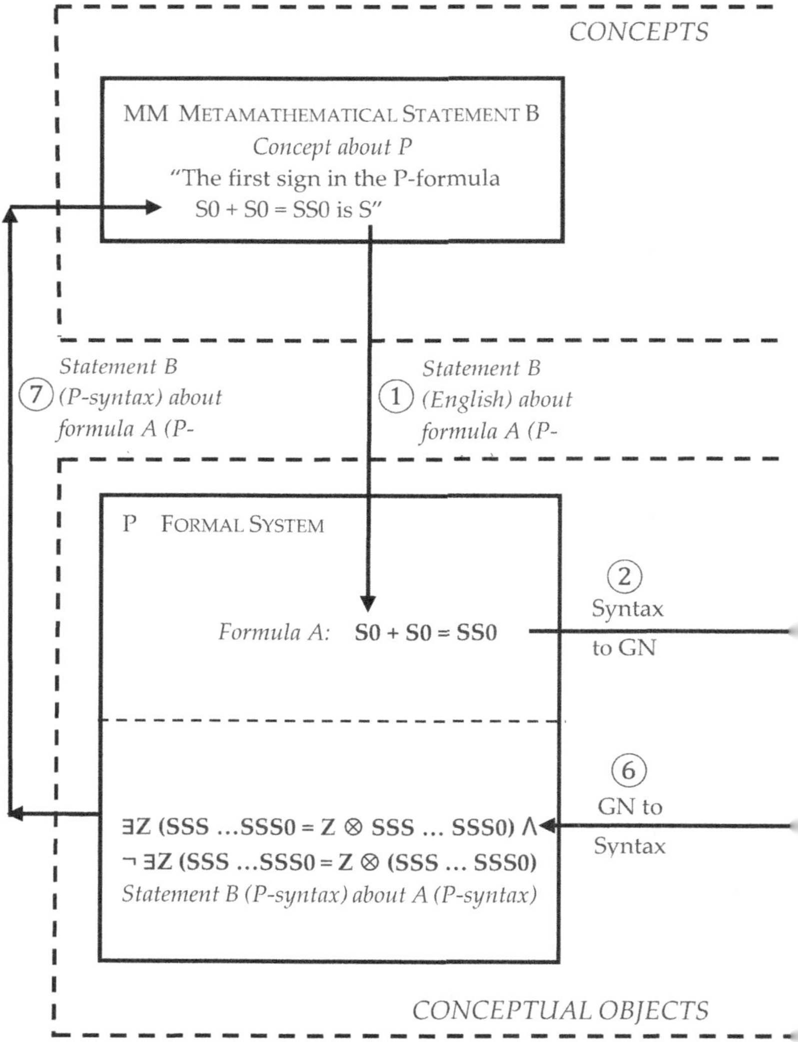

FIGURE 3 - METAMATHEMATICAL SYSTEM, CONCEPT LOOP

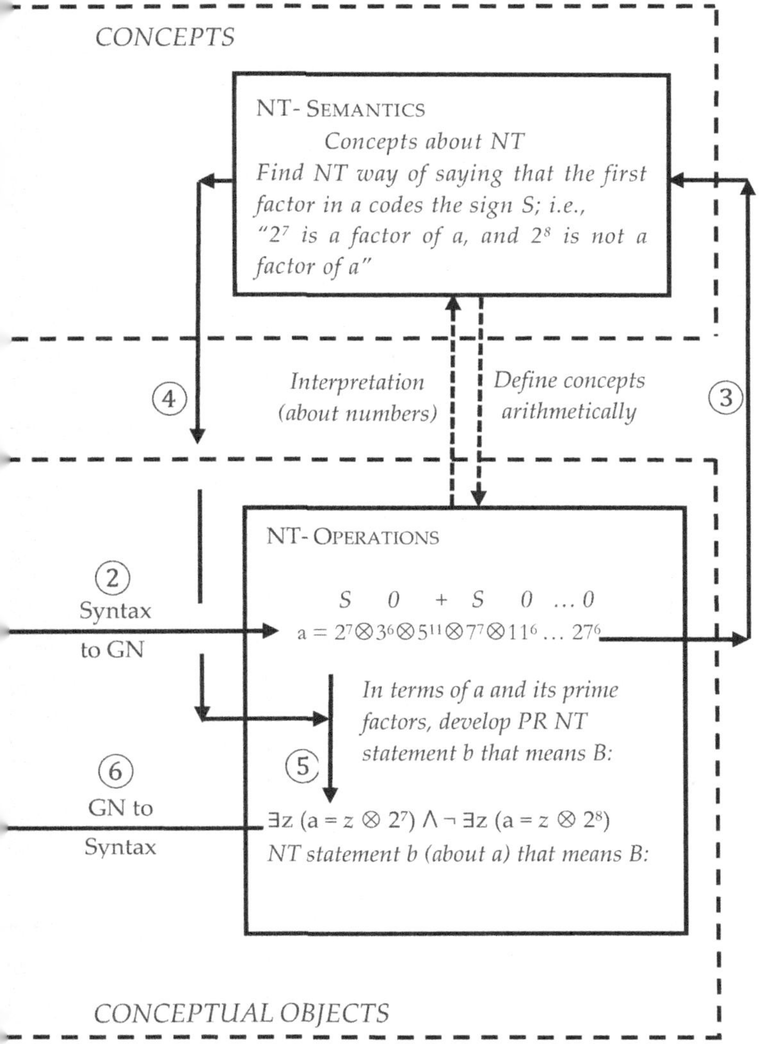

FIGURE 4 – THE CONCEPT OF PROOF

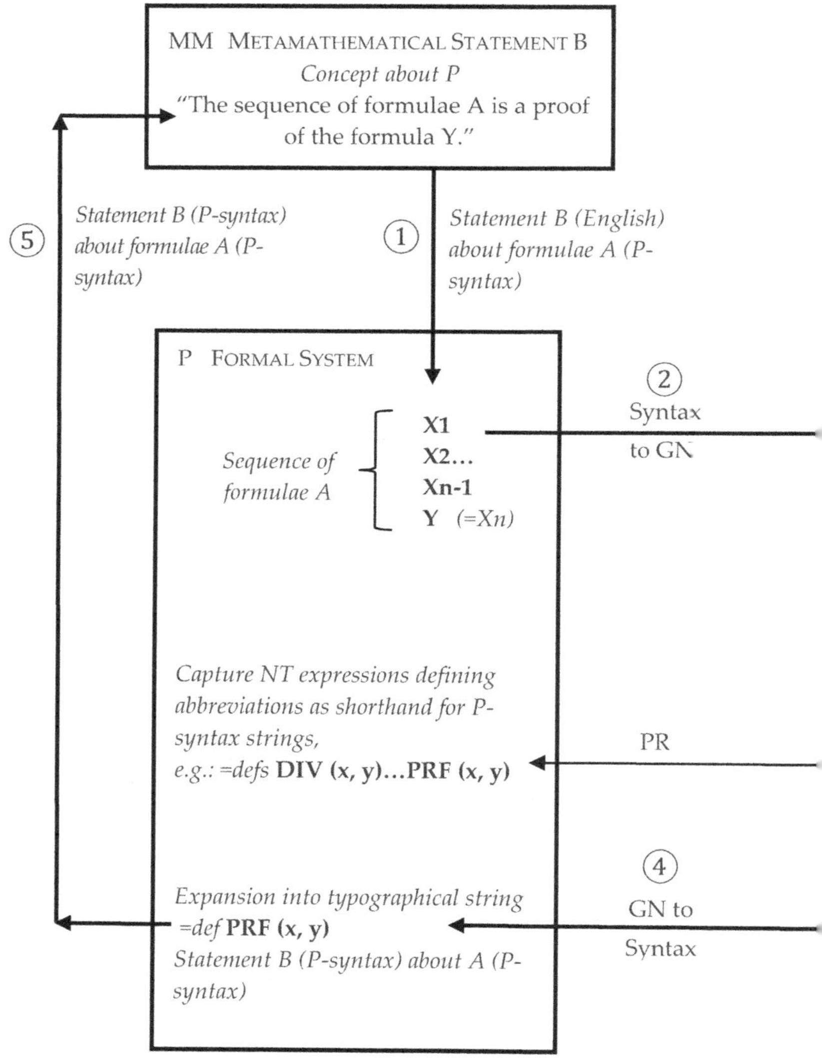

Aristotle
Father of Logic
384 to 322 BC

NT DEDUCTIVE SYSTEM
(Semantics and Operations)

(2) Syntax to GN →

(a =) x; x1, x2…y *(where* x *is the Gödel product of* x1, x2…y)

Find NT way of saying that the sequence of formulae coded by GN = x is a proof of the formula coded by GN = y
i.e., in terms of x, y, and prime factors, develop NT statement b that means B:

(3)

Gödel's defs 1-45

PR

1 Div (x, y) =def

$\exists z\ [(z \leq x) \wedge (x = y.\ z)]$

… L(x)… Gl(x)…

45 Prf (x, y) =def

x ∧ [L(x)] Gl(x) = y

(4) GN to Syntax

b = Prf (x, y)
NT statement b about x, y that means B

Description of Figure 4

Figure 4 illustrates how the concept of proof is developed through the use of *definitions*. In NT, a definition is a name (usually a mnemonic) given to the union of a particular semantic conceptualization and the symbolic object conceptualized; that is, a definition is a named instance of the NT-concept loop shown in Figure 3. For example, the definition of division combines the NT-semantic concept of a relationship (inverse multiplication) between a whole x and a part y, named Div (x, y), together with the NT-symbology $\exists z\ [(z \leq x) \wedge (x = y \otimes z)]$; i.e., Div (x, y) = def $\exists z\ [(z \leq x) \wedge (x = y \otimes z)]$. This example is pertinent, as division was the first in a hierarchy of 46 definitions provided by Gödel in his paper to build up to the definition of provability.

There is an important distinction between the metamathematical statements of Figures 3 and 4. The MM statement of Figure 3 [the first sign in the P-formula S0 + S0 = SS0 is S] is about a *particular* formula A [S0 + S0 = SS0]. The MM statement of Figure 4 [The sequence of formulae A is a proof of the formula Y] is a *general* statement about a sequence of formulae A that is comprised of an unspecified number of formulae Xi, where Xi, Y, and A are variable but can nevertheless be numbered according to Gödel's mapping scheme. We will see below that Figure 5 [Self-referentiality] and Figure 6 [Unprovable Self-referential Statement], like the concept of proof discussed here, refer to general MM statements, whereas Figure 7 [The Gödel Sentence] is about a particular statement—the one that Gödel shows to be undecidable.

Demonstrating in P that Y is the result of the axiomatic method for the general case of A is not feasible because the typographical properties of immensely long chains of P-signs are unmanageable, but the arithmetic properties of prime factorizations of the corresponding Gödel numbers, which are fundamentally rooted in the logic of P, are amenable to the devices of NT.

The metamathematical statement B in English, "The sequence of formulae A is a proof of the formula Y," is captured in P-syntax via a cycle of five pathways as follows:

—Pathways 1, 2; the sequence of formula A in P-syntax is Gödel numbered for processing in NT.

—Pathway 3; exercising intuition and ingenuity, the goal is to find a way of saying in NT that the sequence of formulae coded by GN = x is a proof of the formula coded by GN = y; i.e., to develop an NT statement b that means B in terms of x, y, and prime factors. The mnemonic for the statement b in this case is Prf (x, y), which is Gödel's 45th definition. Prf (x, y) is built up primitive recursively through a progression of NT concepts and mnemonics. Specifically, Prf (x, y) =def x ∧ [L(x)] Gl(x) = y, where L(x) [length of the string x] and Gl(x) [the term x] is developed progressively from concepts originating with 1: Div (x, y) [x is divisible by y], then 2: Prime (x) [x is a prime number]... 45: [x is a proof of y]. (Definition 46 [x is a provable statement] is the only one that is not mathematically primitive recursive, but rather is a metamathematical statement central to Gödel's metamathematical proof as shown in Figures 8 and 9.)

—Pathways 4, 5; the NT statement *b* is then converted into P syntax, producing an MM statement [PRF (x, y)] about A; namely, the sequence of formulae A constitutes a proof.

FIGURE 5 – SELF-REFERENTIALITY

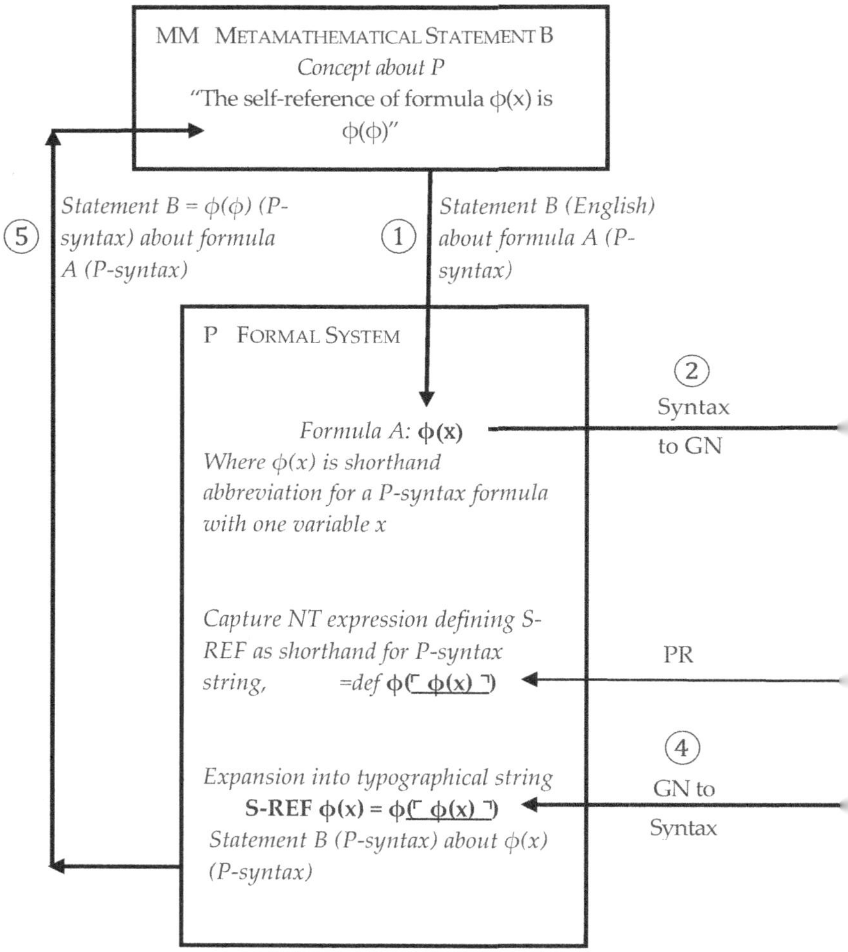

Kurt Gödel
1925

NT DEDUCTIVE SYSTEM
(Semantics and Operations)

② Syntax to GN → *(a =)* ⌜ϕ(x)⌝ *GN of* ϕ*(x)*

Find NT way of saying that the formula coded by GN = ⌜ϕ*(x)*⌝ *is self-referential i.e., in terms of* ⌜ϕ*(x)*⌝ *and prime factors, develop NT statement b that means B:*

PR — *Define* S-ref ϕ(x) *=def* ϕ(⌜ϕ(x)⌝), *i.e., substitute* ⌜ϕ*(x)*⌝ *for x in* ϕ*(x)*

④ GN to Syntax — b = S-ref ϕ(x) = ϕ(⌜ϕ(x)⌝)
NT statement b about ϕ*(x) that means B*

Description of Figure 5

Gödel's proof of the Incompleteness Theorem centers on a self-referential statement, the Gödel Sentence, which he developed indirectly in NT and formulated within the logic of P;[130] see Figure 5. A self-referential formula $\phi(x)$ is one in which a formula ϕ is substituted for its own argument x, thereby yielding the formula $\phi(\phi)$. Importantly, all the formulas in the proof can be defined by primitive recursive functions, which themselves can be defined in first-order Peano arithmetic.

Self-referential statements have caused difficulties in foundational logic, mathematics, and philosophy,[131] dating back to the Greek philosopher Eubulides—another contemporary of Aristotle—who posed the *liar paradox*. The liar paradox may be stated as, "this statement is false." If the statement is true, then it is false, but if the statement is false, then it must be true, and so on. Gödel recognized that definitional circularities lie at the root of self-referential paradoxes.[132] He says in his paper, "The analogy between this result and Richard's antinomy (paradox) leaps to the eye; there is also a close relationship with the 'liar' antinomy."[133] Gödel modified the liar statement about truth to one about provability, which is not paradoxical,[134] "this statement is not provable," further commenting that, "Contrary to appearances, such a proposition involves no faulty circularity, for initially it [only] asserts that a certain well-defined formula (namely, the one obtained from...by a certain substitution) is unprovable. Only subsequently (and so to speak by chance) does it turn out that this formula is precisely the one by which the proposition itself was expressed."[135]

Pathways 1-5 in Figure 5 illustrate how the metamathematical statement B, "The self-reference of formula $\phi(x)$ is $\phi(\phi)$," in English, can be expressed in P-syntax.

—Pathways 1, 2; the formula A = $\phi(x)$ in P-syntax is Gödel numbered for processing in NT. Note that $\phi(x)$ is a *general* statement that maintains x as a variable in the Gödel numbering, which is symbolized by ⌜ ϕ(x) ⌝. The symbols ⌜ and ⌝ are called Quine Quotes, indicating that the included contents ϕ(x) are a Gödel number

—Pathway 3; The goal is to find a way of saying in NT that the formula coded by GN = ⌜ ϕ(x) ⌝ is self-referential; i.e., in terms of ⌜ ϕ(x) ⌝ and prime factors, develop an NT statement b that means B. The mnemonic for the statement b in this case is S-ref, which is Gödel's 30th definition.

—Pathways 4, 5; the NT statement *b* is then converted into P syntax, producing an MM statement [S-REF ϕ(x) = ϕ(⌜ ϕ(x) ⌝)] about A [ϕ(x)]; namely, the formula A is self-referential.

Description of Figure 6

We construct a self-referential formula in Figure 6 that informally says, "I am not provable," and prove that this sentence is neither provable nor disprovable within the theory. Figure 6 merges the concepts of proof and self-referentiality where the concept of proof is qualified by negation; thus, the MM statement B is, "There is *not* a proof of the self-referential formula ϕ(ϕ)." This MM statement in English is established in P-syntax as depicted along five pathways:

—Pathways 1, 2; the formula A = ϕ(ϕ(x)) in P-syntax is Gödel numbered for processing in NT. Note that ϕ(ϕ(x)) is a *general* statement that maintains x as a variable in the Gödel numbering, symbolized by ⌜ ϕ(ϕ(x)) ⌝.

—Pathway 3; The goal is to find a way of saying in NT that the formula coded by GN = ⌜ ϕ(ϕ(x)) ⌝ is not provable; i.e., in terms of ⌜ ϕ(ϕ(x)) ⌝ and prime factors, develop an NT statement b, which we name U, that means B: U(y) [136] = def ∃z(Prf(x; z) ∧ S-ref (y; z)); or, in English, x proves z, where (∧: substitute 'where' for 'and') z is the self-reference of y.

—Pathways 4, 5; the NT statement U is then converted into P syntax, producing an MM statement [¬∃z(PRF(x, z) ∧ S-REF(y, z))] about A; namely, there is not a proof of the self-referential formula A [ϕ(ϕ(x))].

FIGURE 6 - U - UNPROVABLE SELF-REFERENTIAL STATEMENT

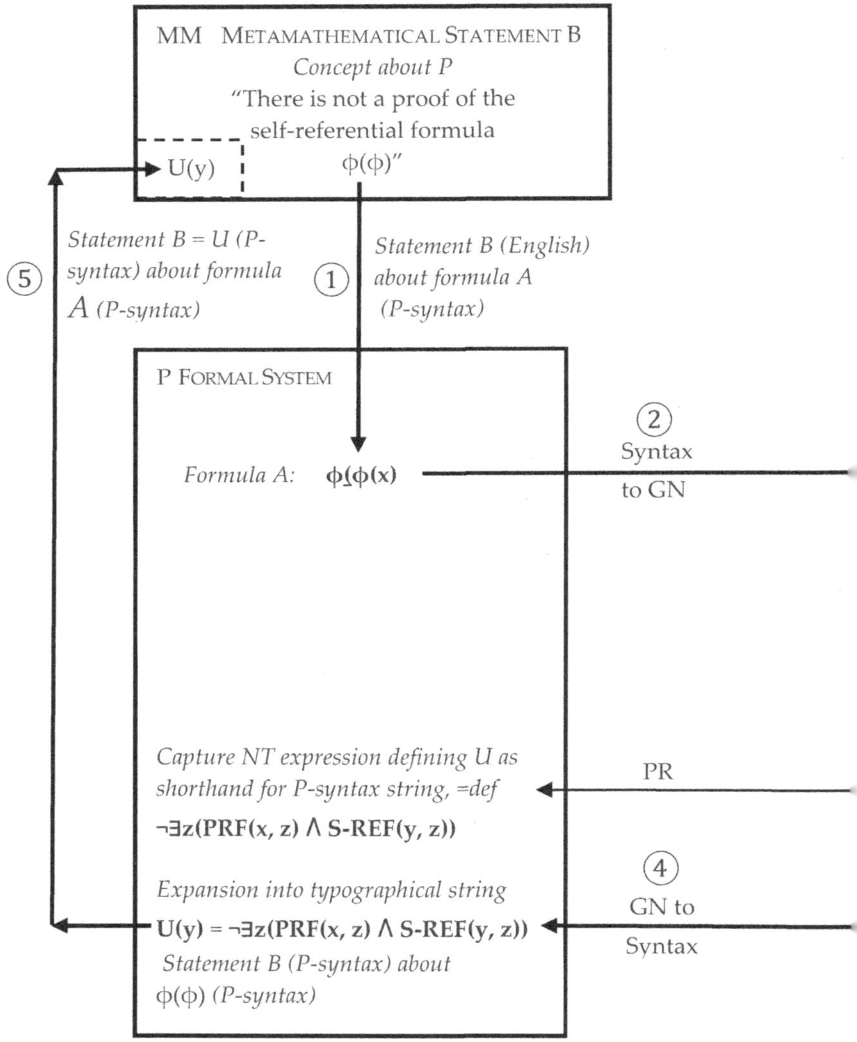

Gödel and Einstein
Best of Friends
1945

NT DEDUCTIVE SYSTEM
(Semantics and Operations)

② Syntax to GN → *(a =)* ⌜ ϕ(ϕ(x)) ⌝

Find NT way of saying that there is not a proof of the formula coded by
GN = ⌜ ϕ(ϕ(x)) ⌝
i.e., in terms of ⌜ ϕ(ϕ(x)) ⌝ *and factors, develop NT statement (b =)* U *that means* B:

③ *Define* U(y)
=def ¬∃z(Prf(x, z) ∧ S-ref(y, z))
PR — *U does not prove z, where z is S-ref of y.*

④ GN to Syntax — U(y) = ¬∃z(Prf(x, z) ∧ S-ref(y, z))
NT statement U about ϕ(ϕ) that means B

FIGURE 7 - G – THE GÖDEL SENTENCE

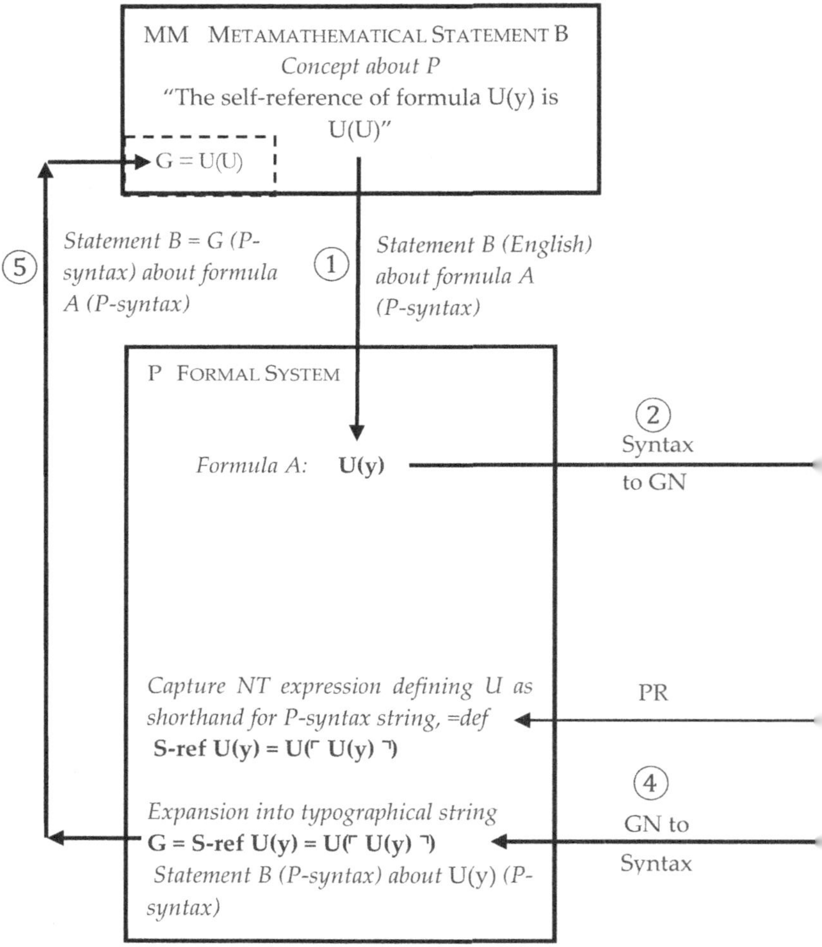

Description of Figure 7

The procedure for developing the Gödel sentence in Figure 7 is the same as that which was applied in Figure 5, except that the formula A is now invariable. The Gödel number for U has a *particular* (constant) value because the variable y in Figure 5 was assigned a specific Gödel number in accordance with the numbering scheme. Thus, we repeat the five pathway steps of Figure 5, replacing the general (variable) formula ϕ(x) with the particular (constant) formula U(y).

FIGURE 7 - G – THE GÖDEL SENTENCE

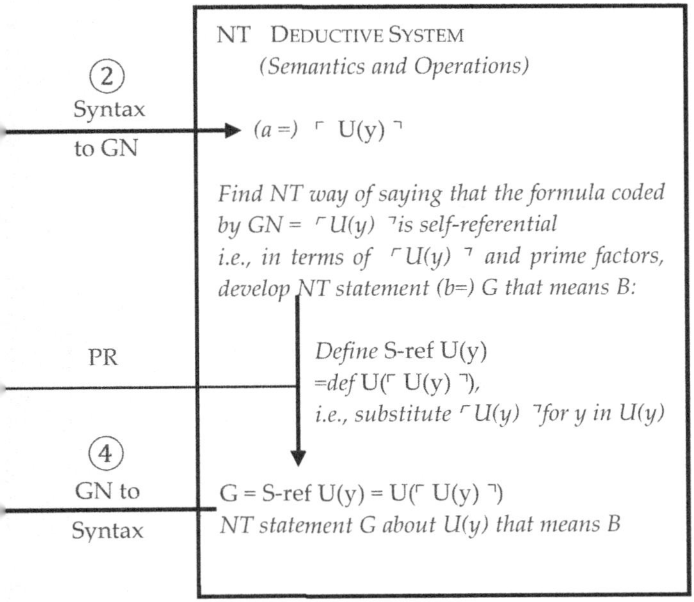

—Pathways 1, 2; the formula A = U(y) in P-syntax is Gödel numbered for processing in NT

—Pathway 3; The goal is to find a way of saying in NT that the formula coded by GN = ⌜ U(y) ⌝is self-referential; i.e., in terms of ⌜ U(y) ⌝ and prime factors, develop NT statement, called G, that

means B: Define S-ref U(y) =def U(⌜ U(y) ⌝), i.e., substitute ⌜ U(y) ⌝ for y in U(y)[137]
—Pathways 4, 5; the NT statement G is then converted into P syntax, producing an MM statement [S-ref U(y) = U(⌜ U(y) ⌝)] about the self-referential formula A [U(y)], which says that G is a self-referential statement that has no proof.

FIGURE 8 - GÖDEL'S PROOF OF THE METAMATHEMATICAL STATEMENT, G IS UNDECIDABLE

PART 1: G IS NOT PROVABLE

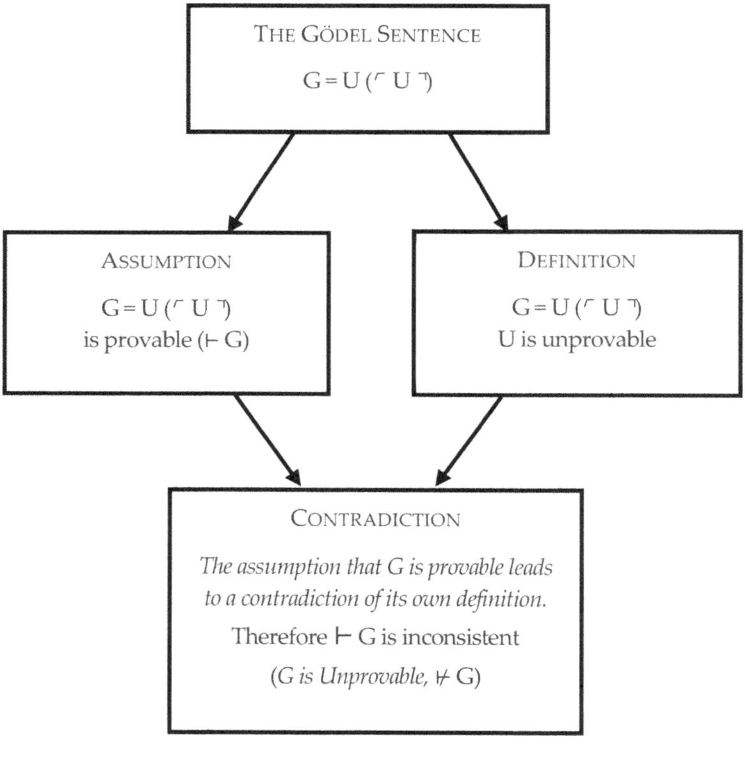

Description of Figures 8 and 9

Figures 8 and 9 involve the concept of *provability*. There is a subtle distinction between the terms "proof" and "provability." Gödel's 45th definition (developed in Figure 4) specifies the conditions for which a sequence of formulae x is a proof of the formula y [Prf (x, y) =def x ∧ [L(x)] Gl(x) = y]. Gödel's 46th definition states that if x is a "proof" of y, then y is

FIGURE 9 - GÖDEL'S PROOF OF THE METAMATHEMATICAL STATEMENT, G IS UNDECIDABLE

PART 2: G IS NOT DISPROVABLE

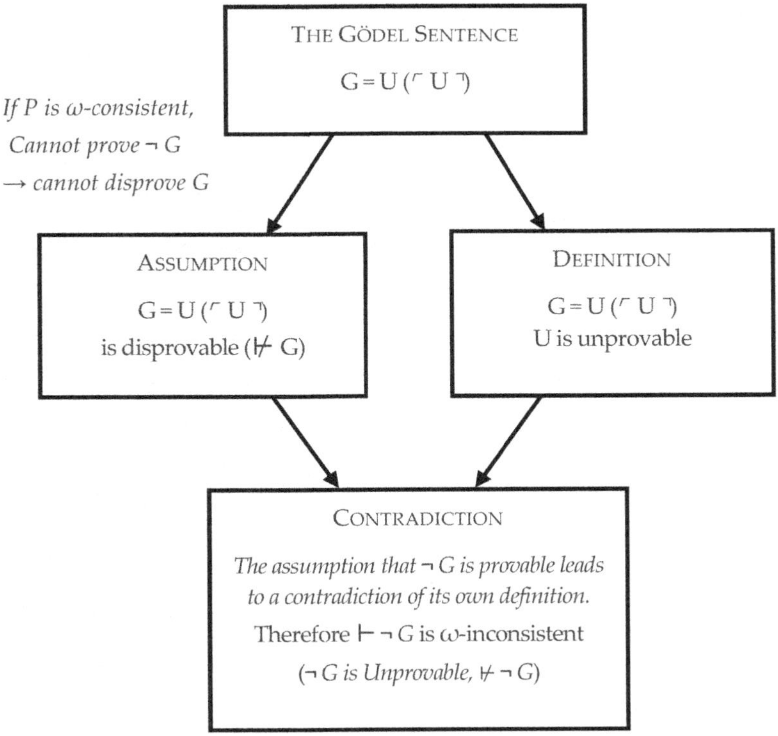

"provable" [Prv (y) =def (∃x) Prf (x, y), where Prv is the mnemonic for "provable"]. Gödel comments in his paper, "Prv is the only one of the concepts 1-46 of which it cannot be asserted that it is recursive." Prv is not built up (primitive) recursively by rules of inference from the axioms of a mathematical system, as is Prf in NT. Prv is a statement *about* Prf, and is therefore a metamathematical concept.[138] Nonetheless, proofs of metamathematical propositions may be derived outside of a mathematical system on the basis of logical principles—which is how Gödel concluded his proof of the Incompleteness Theorem.

The lynchpin of Gödel's metamathematical proof is the logical principle *reductio ad absurdum*—used by Aristotle and Euclid—which asserts that a statement has been proven when it can be established that the assumption of the opposite of that statement leads to a contradiction.

Proof of Part 1:
By *definition*, G = U (⌜ U ⌝) is the self-reference of the unprovable statement U (Figures 6 and 7). As a consequence of G's inherent property of unprovability in ⌜ U ⌝, the *assumption* that G is provable (⊢ G; Figure 8) leads to a contradiction of its own definition, therefore G is not provable.

In more detail:
For conciseness, and granting that the proof x of a self-referential statement encompasses its conclusion y, we abbreviate its symbology [from ∃z(Prf(x, z) ∧ S-ref(y, z))] to ∃xPrfS-ref(x, y) (and abbreviate a self-referential statement that has no proof [from ¬∃z(Prf(x, z) ∧ S-ref(y, z))] to ¬∃xPrfS-ref(x, y)).

- G is by definition the S-ref of the formula U =def ¬∃xPrfS-ref(x, y), i.e., G is the formula ¬∃xPrfS-ref(x, ⌜ U ⌝)
- *Assume* ⊢ G (G is provable); i.e., (i) ⊢ ¬∃x(PrfS-ref(x, ⌜ U ⌝). If G is provable, there exists *some* GN m that codes its proof.
- But G is by *definition* the S-ref of the formula with GN ⌜ U ⌝, so that ⊢ PrfS-ref(m, ⌜ U ⌝); i.e., m indirectly codes a proof of U.

- Existentially quantifying for m yields (ii) $\vdash \exists x PrfS\text{-}ref(x, \ulcorner U \urcorner)$.
- Now, expressions (i) and (ii) are contradictory and cannot both be correct if P is a consistent system; the assumption $\vdash$ G leads to an inconsistency.
- Therefore, if P is consistent, $\nvdash$ G.

It was demonstrated in Figure 8 that the Gödel sentence G is not provable, $\nvdash$ G. Figure 9 demonstrates that G is also not disprovable; that is, the negation of G is not provable, $\nvdash \neg$ G. If G and $\neg$ G are both unprovable, then, by definition, G is undecidable. The proof that $\neg$ G is unprovable follows the same line of reasoning as the proof that G is unprovable, *reductio ad absurdum*. However, Gödel's proof that $\nvdash \neg$ G warrants clarification of two relevant issues: *ω–consistency* and *proving a negative.*

ω–consistency

Gödel's proof that $\nvdash \neg$ G employs a generalizing form of consistency, called *ω–consistency* (ω is a conventional label for the set of natural numbers), which accommodates extended interpretations of P. An ω-consistent system is able to prove statements concerning not only the finite natural numbers, but also statements about infinite sets and non-standard numbers.[139] ω–consistency is stronger than standard, "finitist," consistency[ix] which concerns proofs and other logical procedure that can be completed in a finite number of steps. Gödel proved that if $\neg$ G is assumed to be provable then P violates ω-consistency.

---

ix The elements of all interpretations of P (and all sufficiently strong systems, including Peano Arithmetic) are linearly ordered and possess an initial segment isomorphic to the standard natural numbers (1, 2, 3…). An extended, non-standard interpretation is one that has additional elements (x) outside this initial segment. An ω–consistent system cannot prove *both* (the negation) formula $\neg \phi(n)$ for each natural number: $\neg \phi(0), \neg \phi(1), \neg \phi(2)$… *and* $\phi(x)$ for some (non-standard) number x; or that all of $\phi(0), \phi(1), \phi(2)$… and also $(\exists x) \neg \phi(x)$ are theorems.

Proving a Negative

It is widely believed that "you can't prove a negative." For example, "you can't prove that unicorns don't exist." Even with the claim, based on inductive reasoning, that unicorns have never been observed and (convincingly) documented, how can one know with absolute certainty that unicorns are not hiding somewhere? Further, the assertion about the non-existence of unicorns is not a *proof* based on deductive reasoning; it is rather a just negative statement. In fact, the statement, "you can't prove a negative" is, itself, just a negative statement.

In logic, a negative statement can be proven. Earlier, we described Theophrastus' argument form *modus ponens*:

$A \rightarrow B, A, \therefore B.$

If A then B, A is the case, therefore B is the case.

If it is a robin then it is a bird, it is a robin, therefore it is a bird.

Theophrastus also postulated a complementary argument form termed *modus tollens*:

$A \rightarrow B, \neg B, \therefore \neg A.$

If A then B, B is not the case, therefore A is not the case.

If it is a robin then it is a bird, it is not a bird, therefore it is not a robin.

In this example of *modus tollens,* the conclusion is a proven negative statement.

Proof of Part 2:

By *definition*, $G = U(\ulcorner U \urcorner)$ is the self-reference of the unprovable statement U (Figures 6 and 7). As a consequence of G's inherent property of unprovability in $\ulcorner U \urcorner$, the *assumption* that G is disprovable (i.e., the negation of G is not provable, $\nvdash \neg G$; Figure 9) leads to a contradiction of its own definition, therefore $\neg G$ is not provable.

In more detail:

Again, for conciseness, and granting that the proof x of a self-referential statement encompasses its conclusion y, we abbreviate its symbology [from $\exists z(\mathrm{Prf}(x, z) \wedge \text{S-ref}(y, z))$] to $\exists x\text{PrfS-ref}(x, y)$ (and abbreviate a self-referential statement that has no proof [from $\neg\exists z(Prf(x, z) \wedge S\text{-}ref(y, z))$] to $\neg\exists x PrfS\text{-}ref(x, y)$).

- G is by definition the S-ref of the formula U =def $\neg\exists x PrfS\text{-}ref(x, y)$, i.e., G is the formula $\neg\exists x PrfS\text{-}ref(x, \ulcorner U \urcorner)$
- *Assume* $\vdash \neg$ G (G is disprovable); i.e., (i) $\vdash \neg [\neg\exists x(PrfS\text{-}ref(x, \ulcorner U \urcorner)] = \vdash \exists x(PrfS\text{-}ref(x, \ulcorner U \urcorner)$. If G is disprovable, there exists *no* GN m that codes its proof.
- But G is by *definition* the S-ref of the formula with GN $\ulcorner U \urcorner$, so that $\nvdash PrfS\text{-}ref(m, \ulcorner U \urcorner)$; i.e., m does not code a proof of U.
- Existentially quantifying for m yields (ii) $\nvdash \exists x PrfS\text{-}ref(x, \ulcorner U \urcorner)$.
- Now expressions (i) and (ii) are contradictory and cannot both be correct if P is an ω-consistent system; so the assumption $\vdash \neg$ G leads to an ω-inconsistency.
- Therefore, if P is ω-consistent, $\nvdash \neg$ G.

Figure 8 demonstrates that if P is consistent then the Gödel sentence G is not provable ($\nvdash$ G), and Figure 9 demonstrates that if P is ω-consistent then G is not disprovable ($\nvdash \neg$ G). Hence, assuming ω-consistency, and thus consistency, G is undecidable. Therefore P is incomplete.

Quod Erat Demonstrandum

* * *

12
Artificial Consciousness

Artificial Intelligence and Artificial Consciousness

Artificial Intelligence (AI) has revolutionized the ways in which people access information, solve problems, and communicate. In the past few years, a growing list of advances in Artificial Intelligence has led to improvements in healthcare, education, finance, social media, automation and robotics, cyber security, and numerous other applications. In particular, the natural language interface between people and machines made possible by the transformer architecture has made the capabilities and potential of AI readily available to all.

The future of AI is extremely promising. At the time of this writing in 2026, the volume of data dedicated to AI is doubling every two years, the growth of AI processing speed is even more explosive, doubling every six months, and the creativity of people developing AI applications and systems innovations is inspiring.

Recent developments driving the progress of AI include:
Machine learning—AI computer systems and devices are learning and improving from experience, without being explicitly programmed.
Perception—AI systems are able to use sensory detectors to perceive and learn about the world around them, detecting and processing visual, tactile, olfactory, and audible information. AI systems can take in information far exceeding human capabilities such as detection of sound and light frequencies well beyond the human range.
Semantics—AI systems can capture and apply the meanings of words and sentences by which people understand.

Generative modeling—AI systems can create original content, rather than simply copying and manipulating data.
Autonomy—AI systems can make decisions independent of human intervention to solve problems and achieve goals.

Artificial intelligence is based on a new set of paradigms in computer technology. Traditionally, computers were programmed, by people, with an explicit set of instructions directing them how to accomplish specific tasks. In contrast, AI computer systems are trained to think as people do, learning how to select and perform tasks on their own. They simulate human intelligence. But will it ever be possible for non-living computer systems to be conscious? That is the subject of this chapter. To be sure, Artificial Intelligence has a long way to go before non-living hardware and software systems can possibly attain the heights of human mental consciousness.

This chapter is divided into four sections:
1- History and Future of Artificial Intelligence
2- Artificial Neural Networks, Machine Learning and Training
3- Biological Neural Networks and Mental Consciousness
4- Artificial Consciousness

1- History and Future of Artificial Intelligence

1943 - Neurophysiologist Warren McCulloch and logician Walter Pitts collaborated on a groundbreaking paper titled, *A Logical Calculus of the Ideas Immanent in Nervous Activity*. The purpose of this work was to investigate the possibility of representing logical functions through networks of artificial neurons. They proposed the first mathematical model of a neural network, laying the foundation for AI.[140] The McCulloch–Pitts neural network mimicked the architecture of biological neural networks, but signaling by the artificial neurons was limited to Boolean logic (0s and 1s only), whereas signaling in a biological neural network spans a continuum of real number values.

1948 - William Grey Walter, neuroscientist and roboticist, pioneered the development of the first *electronic autonomous robots* known as "tortoises," named Elmer and Elsie. Equipped with light sensors, touch sensors, and basic control mechanisms, these robots possessed the ability to sense and dynamically respond to their environment.

1949 - Donald Hebb wrote, *The Organization of Behavior,* a work that pointed out how biological neural pathways are strengthened and remodeled through repeated activation, a concept fundamental to the way in which humans learn. Hebb famously said, describing his theory of synaptic plasticity, "Cells that fire together, wire together." This principle is now a central characteristic of artificial neural networks.

1950 – Turing test; in 1950, Alan Turing considered whether machines could possibly think. In his paper, *Computing Machinery and Intelligence,*[141] he claimed there is no convincing argument that machines (digital computers) cannot think intelligently, as humans do; that is, an interrogator could not distinguish between a machine and a human.

1952 - The first *machine learning* program was written by Arthur Samuel to play checkers against itself and improve its strategy by accumulating a database of winning moves and positions. Samuel coined the term "machine learning."

1956 - John McCarthy (the "father of Artificial Intelligence"), alongside influential figures including Marvin Minsky (co-founder of MIT's AI laboratory) and Claude Shannon (the "father of information theory"), orchestrated a workshop held at Dartmouth College which is considered the event that kicked off Artificial Intelligence as a research discipline. McCarthy coined the term "Artificial Intelligence" at this meeting.

1957 - The first *trainable neural network,* the Perceptron, was demonstrated by psychologist Frank Rosenblatt. The Perceptron was a feedforward neural network comprised of 3 layers: an input layer, one intermediate layer with randomized weights (Hebbian strengths) that did not learn, and an output layer that did learn. *Feedforward n*eural networks were the first and most basic artificial neural network configurations, dating back to the McCullouch-Pitts architecture. Subsequent types of artificial neural networks include: *Convolutional*—a feedforward neural network using filters and pooling layers to implement an orderly procedure whereby sources of information are intertwined; *Recurrent*—results are recycled back into the network, availing network memory; and the *Transformer* architecture— employing massive parallel processing and an attention mechanism.

1965 - The first working *deep learning* algorithm, introduced by Alexey G. Ivakhnenko, was the "Group Method of Data Handling." This was a method for training arbitrarily deep networks (having multiple "hidden" layers between the input and output layers). Deep learning is a type of machine learning that utilizes deep networks to learn non-linear relationships, necessary to model the complex decision-making power of the human brain.

1966 – ELIZA was an early *natural language processing* computer program developed by Joseph Weizenbaum that could respond to a human using a natural language. Eliza may be considered the first *chatbot* (a computer program that engages in conversation with humans), and also the first example of *generative AI* (AI that can create new content such as images, text, audio, music, and videos).

1970 - The earliest published instance of the modern form of *back-propagation* (a machine learning technique that works backward from outputs to inputs by adjusting intermediate weights) was presented by Seppo Linnainmaa. Backpropagation would become essential for the optimization of artificial neural networks.

1971 - The first commercially produced *microprocessor,* the Intel 4004 central processing unit (CPU), was released. By the late 1970s, building on microprocessor technology, the potential of AI would be unleashed.

1972 - MYCIN, an early *expert system* designed to diagnose and prescribe medications for blood diseases was unveiled. Expert systems are AI programs that simulate logical human reasoning and decision-making in a specific domain. The concept of expert systems was developed by computer scientist Edward Feigenbaum at Stanford University.

1975 - Paul Werbos created the backpropagation algorithm to improve the training accuracy of neural networks.

1978 – The XCON expert learning system was a production-rule-based system written by John P. McDermott of Carnegie Mellon University, marking the commencement of the golden age of expert systems. In the 1980s, expert systems were adopted by corporations around the world, and knowledge systems became a mainstream focus of AI research.

1983 – The *Internet* was created by ARPANET (Advanced Research Projects Agency Network).

1986 - The training of deep neural networks was advanced by David Rumelhart, Geoffrey Hinton (the "godfather of deep learning," and Nobel laureate in Physics), and Ronald Williams, in their seminal paper, *Learning representations by back-propagating errors.*[142]

1990 - The first practical application of backpropagation came about through the work of Yann LeCun at Bell Labs. He used backpropagation in combination with a convolutional neural network to classify handwritten digits. Backpropagation is an algorithm used to alter and optimize the weights of neurons throughout a network, generating a form of internal memory; as opposed to a recurrent neural network in which neurons with established weights feed back into themselves.

1991 – The first *website* was designed by Tim Berners-Lee while working at CERN, the particle physics laboratory in Switzerland. He used hypertext, which is an address tag attached to a word or phrase, to seamlessly access documents on different computers connected to the Internet (previously, one had to log onto each computer individually to access its data). He termed the system the "Worldwide Web." Thereafter, the Web made available a wealth of data for the training of AI models.

1997 – Deep Blue, IBM's supercomputer, defeated Gary Kasparov, the reigning world chess champion. Deep Blue was an expert system trained with only the rules of chess.

2005 – The DARPA (Defense Advanced Research Projects Agency) grand challenge was a landmark in the field of robotics. The Stanford Racing Team's unmanned vehicle, "Stanley," won DARPA's $2 million prize. The autonomous vehicle traversed a 132-mile course consisting of rugged terrain, open desert, tunnels, and Beer Bottle Pass (an infamous winding mountain pass with a sheer drop on one side, and a rock face on the other).

2011 – IBM's Watson, a group of software tools running on a massively parallel supercomputer, was designed to comprehend natural language and answer questions. Watson won the TV quiz show Jeopardy by soundly defeating the two all-time champions, Brad Rutter and Ken Jennings. Watson executed hundreds of algorithms simultaneously to analyze a question from many perspectives, rate potential answers, and gather additional validation for each answer.

2012 –AlexNet, a powerful visual-recognition network with 650,000 neurons and 60 million parameters was designed at the University of Toronto by Alex Krizhevsky, Ilya Sutskever, and Geoffrey Hinton. AlexNet dominated the ImageNet Large Scale Visual Recognition challenge by achieving 85% accuracy in classifying objects.

2013 – Word2vec was released by Google. In this deep learning neural network, inputs, weights, and outputs are represented mathematically by vectors. For example, in natural language processing, words are transformed into vectors such that words with similar *meanings* are positioned "close" to each other in the vector space. These vectors are thus endowed with semantic and syntactic information—"A word is characterized by the company it keeps." [143] In general, by organizing data into vectors, machine learning models can represent various features and operations on the data, such as clustering, classification, and discovering patterns within a data set. Other vector applications include image recognition and recommendation systems.

2014 - Alexa, a voice-controlled *virtual assistant* was released by Amazon. Alexa could search the internet, play music, control smart devices, create lists, shop, translate audio and video calls in real time, and so forth. Also in 2014, the *Generative Adversarial Network* (GAN) was developed by Ian Goodfellow and colleagues, in which two deep learning neural networks competed against each other to mutually refine and enhance their capabilities.

2015 - Improvements in computing speed and cost, coupled with the growth of "big data," significantly advanced machine learning. Computers could now intake and process massive data sets, and, at this time, Google had more than 2,700 AI projects in development.[144]

2016 - AlphaGo, using reinforcement learning and self-play (not GAN) was developed by DeepMind. The program was taught only the rules of the game, and created strategy by itself sufficient to defeat legendary Go player Lee Sedol with an inventive and unlikely move that overturned centuries of traditional wisdom.

2017 - The *Transformer* neural network architecture was introduced in the groundbreaking paper, *Attention is All You Need*, by Ashish Vaswani et al, of the Google Brain team, which stated, "We propose a new simple network architecture, the Transformer, based solely on *attention* mechanisms, dispensing with recurrence and convolutions entirely." The transformer architecture has become the foundation of many modern language models and other generative AI applications.

2020 – GPT-3 (*Generative Pre-trained Transformer*), with 175 billion parameters, was released by OpenAI, establishing itself as one of the most powerful natural language models of its era. Based on input prompts, GPT-3 generated coherent, contextually relevant human-like speech and text.

2022 – ChatGPT, released by OpenAI, was a Chabot that marked a pivotal milestone in the evolution of AI. It could answer questions, compose essays, create poetry, and engage in conversations.

2023 – AutoGen, an *agentic AI* system, was released by Microsoft. AI agents are systems that work autonomously, following general instructtions to use a variety of tools to complete tasks. Whereas generative AI creates new content based on learning from training data, agentic systems not only generate content, but also take actions based on information *they* gather from their environment. For example, ChatGPT requires human input to elicit a response. In contrast, autonomous agents are not directly invoked for a task, but activate themselves. They lie in wait, monitoring conditions until they determine when to self-activate. They anticipate needs and provide proactive support, such as sending appointment reminders, or notifying customers about potential issues before they arise. (The idea of agent-based systems was developed in the 1940s, but due to limited computing power did not become practicable until the 1990s. The 1990s also saw the introduction of multi-agent systems, in which multiple AI agents collaborate to achieve goals too complex for a single agent to handle. This approach opened new avenues for applications in robotics, simulations, and complex problem solving—all of which have been realized today.)

2026 and beyond -

The following projections for the future of AI are based upon ongoing business ventures, as well as research and development programs currently underway at the time of this writing in 2026. Creative ideas and discoveries will undoubtedly continue to emerge, as they have throughout the history of technology, expanding the potential and fulfilling the destiny of AI.

Multimodal AI - AI products will continue to increase, improve, and integrate information from different modalities, including text, speech, images, video, audio, numerical data, programming code, and AI generative sources, ultimately surpassing human capabilities of sensing, memory, learning, language, attention, and thinking.

Balanced AI Infrastructure – Certain AI applications will migrate from today's exclusively large, closed models to smaller, open, more versatile AI solutions that offer end-users: affordability and efficiency, tools to create custom agents with new and novel behaviors, and community collaboration in "federated" AI projects. While large back-end systems will continue to provide computational power and economies of scale, the AI infrastructure of the future will take advantage of both centralized and distributed resources.

Internet of Things (IoT) - A network of physical devices, vehicles, appliances, and other physical objects embedded with sensors, software, and network connectivity, will allow them to collect and share information and functionalities, leading to smarter homes, industrial systems, cities, as well as a more secure and scalable AI ecosystem. Smartphones, IoT gadgets, and "edge" (local) computing nodes will collaborate and share insights without transmitting raw data.[145]

Neuromorphic computing - Neuromorphic computing is a specialized approach to AI computing inspired by the brain's structure and operation. It uses analog circuitry to mimic the way the human brain processes information, whereas traditional digital computing does not. Neuromorphic computing will demand low energy consumption, be ideally configured for applications like IoT and edge computing, and serve as a building block of artificial superintelligence.[146]

Optical computing – Leveraging the fact that photons travel significantly faster than electrons, the speed advantages of optical computing will provide computational acceleration, improved data transfer, and energy efficiency gains, making it a good fit for the demands of AI. Optical computing finds existing applications in high-speed data transmission through fiber optics, optical interconnects between electronic chips, and image processing. Optical computing research and development is ongoing for parallel processing, quantum computing, and machine learning applications.[147]

Quantum AI - Using the unique properties of qubits (quantum bits) to represent AI data will mitigate the computational constraints of classical computers that process only binary data; i.e., information encoded by strings of two possible values, typically denoted by 0s and 1s. Rather than processing binary data strings sequentially, as classical computers do, quantum computers process *superpositions* that enfold all possible states of the data simultaneously. Quantum computing research and experimentation is underway in cryptography, cybersecurity, financial modeling, drug development, weather forecasting, and optical neural networks for machine learning.[148]

The Singularity - A hypothetical point in the future where AI surpasses human intelligence and begins to improve itself autonomously and exponentially. Future AI machines may have ideas very different from those of human beings regarding how they should evolve and what decisions will be made.[149]

Consciousness - Machines become so sophisticated that they are self-aware and *experience* reality in the same way as humans.[150, 151] Is it possible, through autonomous learning and self-development, for information integrating AI machines[152] to reach a level of complexity sufficient to attain holistic closure, together with an emergent *ultra-sensory* neural network, as occurred in the evolution of the biological neural network of the human brain? That is, can a machine attain mental consciousness?

2- Artificial Neural Networks, Machine Learning and Training

Artificial Intelligence is a branch of computer science concerned with "machines" (hardware and software) that simulate the aptitudes of human intelligence, such as learning, reasoning,[153] perception, and language comprehension. In AI, these aptitudes are realized through the use of *models*—collections of *algorithms* linked together to achieve desired objectives; a model is an algorithm applied to certain kinds of data. Algorithms, in turn, are the instructions—a list of procedures or a flowchart—that outline the logic to be encoded in a software program.

AI models are designed to run on *artificial neural networks*.[154] Although there are a variety of AI platforms, such as the symbolic architectures,[155] we will focus on "connectionist" artificial neural networks, as they are the predominant architecture today, and most closely resemble the *biological neural networks* underlying human mental consciousness.)

An artificial neural network is a software simulation of the human brain's biological neural network. The basic component of the biological neural network is the biological neuron (Figure 12-1). Biological neurons convey signals by emitting and reacting to electrochemical ions (neurotransmitters) that traverse tiny clefts (synapses) between the output end (axon) of one neuron and the input end (dendrite) of another. Upon the accumulation of a specific voltage (action potential firing threshold) on the axon hillock of the receiving neuron, it, in turn, is triggered to release neurotransmitters from its output end into synapses that connect to other neurons in the network.

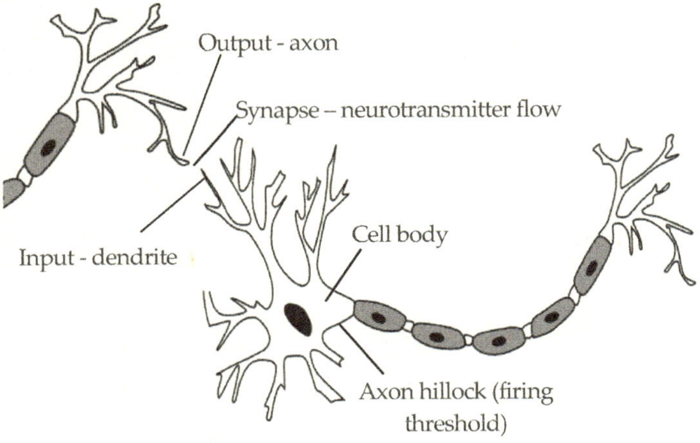

Figure 12-1 *Biological Neuron*

The basic component of an artificial neural network is the *artificial neuron* (Figure 12-2). An artificial neuron is a mathematical function designed to process data and transmit it to other neurons in the network. Like biological neurons, artificial neurons transmit data through a hierarchy of layers, from the input layer to the output layer (Figure 12-3) which illustrates a basic feedforward network containing three intermediate, or "hidden" layers.

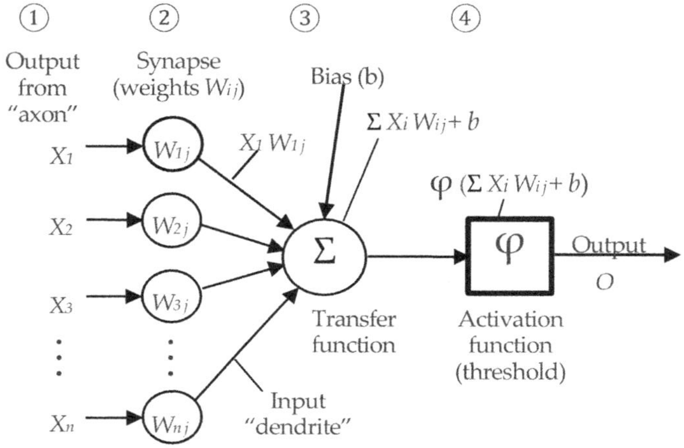

Figure 12-2 *Artificial Neuron*

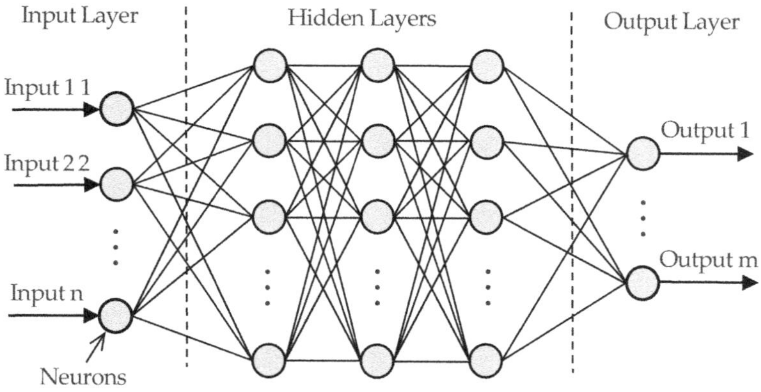

Figure 12-3 *Artificial Neural Network*

The flow and processing of data for an artificial neuron consists of four steps:
Step ① pertains to the neuron's initial source of data, X_i. For neurons in the first layer of a network, the source of data is a dataset, a collection of information coming from text documents, audible speech and visual images (the computer is equipped with sensory mechanisms), spreadsheets and databases, the internet, and so forth. In subsequent layers, the hidden and output layers, the source of data is the output from neurons in the preceding layer.

Step ② is the *synaptic connection point*, where each source datum, X_i, is multiplied by a weighting factor, W_i, yielding the product $X_i W_i$, which determines the significance of the connection. Multiple synapses can connect to a single neuron, with each synapse having a different level of influence on the neuron's process. Synaptic weighting is crucial to network training and learning. As data is refined during the repetitive training cycles, synaptic weights change, and this is how learning occurs.

Step ③ is the linear *transfer function*, Σ, which summates the weighted data values, $X_i W_i$, together with a *bias* factor value, b. The bias factor is introduced to selectively compensate for incomplete data, erroneous assumptions, human prejudices, etc. The summation yields the value, $\Sigma X_i W_{ij} + b$.

Step ④ is the *activation function*, $\varphi\ (\Sigma X_i W_{ij} + b)$, whose output, O, produces various effects on neurons in the next layer. Activation functions are the backbone of neural networks, enabling them to capture non-linear relationships in data, which are necessary for dealing with complexity. Without non-linearity, even networks with multiple hidden layers, i.e., "deep networks," would be limited to solving only simple, linearly separable problems.

ARCHITECTURE OF AN ARTIFICIAL NEURAL NETWORK

Artificial neural networks are designed to accomplish particular objectives such as pattern recognition, data classification, problem solving, making predictions and decisions, comprehending and communicating in natural languages, and much more. Each of these objectives entails a different network design, or "model," but all are developed by employing: 1) a specific method of machine training and learning, 2) an activation function common to all neurons in a given layer, and 3) a distinctive kind of network structure. Examples of these three network design components are listed in the following table:

Components of Network Design

Learning Methods	*Activation Functions*	*Network Structure*
—Supervised learning	—Sigmoid	—Feedforward
—Unsupervised learning	—Tanh	—Recurrent
—Reinforcement learning	—Rectified linear unit	—Convolutional
—Neuroevolution	—Softmax	—Transformer

LEARNING METHODS

Machine learning enables machines to simulate the way humans learn from data such they can be capable of performing tasks without being explicitly programmed. Machine learning is a subset of artificial intelligence concerned with how a computer system *develops* in order to better recognize patterns, make predictions, render decisions, and determine what and when actions will be taken.

Machine training is an essential part of machine learning. Machine training involves the iterative adjustment of internal network parameters, such as weights and biases, until output becomes verifiably reliable. Machine training also requires the selection of *hyperparameters*, i.e., system setup parameters that are not learned from the data. Hyperparameters include the number of hidden layers, number of neurons per layer, activation functions, and network structure.

- *Supervised learning (from data content)*

Supervised learning is a method of machine learning by which the algorithm is trained on an input dataset to produce already known output results (that can be "labeled"). Once successfully trained on particular types of data (or "features"), the algorithm becomes a model which can then be fed new data to produce reliable results. Initially, the algorithm to be trained is furnished with a set of synaptic weights that, by design, does not produce correct results with respect to the known, labeled values, and therefore weights (and biases) must be updated throughout the network to reduce the discrepancy, or cost, between the values of the input features and those of the labeled output ("cost" refers to the aggregate discrepancy over an entire training set, while "error" or "loss" applies to a single training example; we will use the term "error" gener-ically except where context directly refers to "cost"). Adjusting and fine-tuning synaptic weights to reduce error/cost is accomplished through the process of *backpropagation.*

Backpropagation works by feeding output back through the layers of the network, calculating the error, and adjusting the weights to minimize the error (Figure 12-4). A standard procedure for minimizing error is *gradient descent*. Gradient descent works by iteratively adjusting the weights of the network in the direction of the negative gradient of the error function until the minimum (difference between the known output value and the calculated output value) is reached. Gradient descent involves the application of two basic procedures from differential calculus: finding the minimum of a multidimensional, multi-feature function, and applying the chain rule across layers (where weights are modified by a particular activation function per layer).

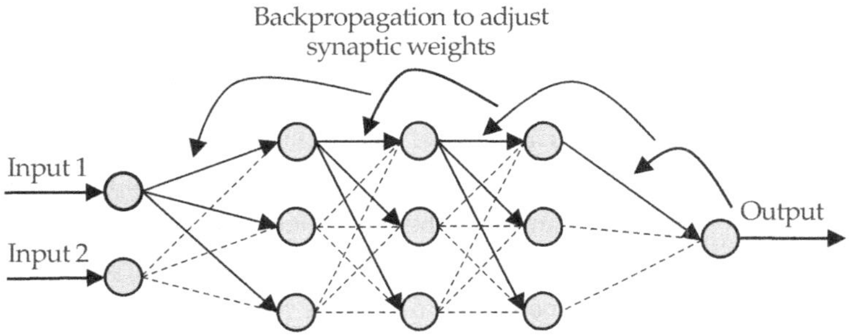

Figure 12-4 *Backpropagation*

Supervised Learning Applications:

Predictive analysis; recommendation engines; image, speech, face, and object detection, recognition, and classification (by identification); fraud and spam detection; public sentiment analysis; regression (the tendency of extreme data values to "revert" to the overall mean value based on a known relationship between independent and dependent variables); quality control; medical diagnoses; natural language processing, analysis, and translation

- *Unsupervised learning (from data structure)*

Unsupervised machine learning involves training a model to discover hidden patterns, structures, and relationships within a dataset without any predefined output or explicit instructions on what to search for.

Unsupervised learning reveals insights that might not be apparent from a labeled dataset.

There are three main approaches to implement unsupervised learning:
—*Clustering* is a technique to group unlabeled data into categories based on their similarities or differences.
—*Association* employs an established set of rules, such as predefined guidelines set by experts or rules of deductive logic, to discover relationships between variables in a dataset.
—*Dimensionality reduction* is a technique used when the number of features, or dimensions, in a dataset is too high. It reduces the number of data inputs to a manageable size while also maintaining the integrity of the dataset. Dimensionality reduction can obtain the benefits of backpropagation network tuning for unsupervised learning projects in an indirect fashion. Backpropagation is usually considered a supervised learning algorithm because initially there must be a known, labeled, output for each input value in order to calculate the error/cost function gradient. In unsupervised learning, input data is compressed by an *autoencoder* sub-neural network into a smaller, *dimensionality reduced* representation known as the latent space which retains the essential features of the dataset. The original input is then reconstructed using the latent space by a *decoder* network. Updating the weights and biases to minimize the reconstruction error between input data and the output of the decoder network delivers the benefits of backpropagation to unsupervised learning applications.

Unsupervised Learning Applications:

Data clustering for classification of objects, patterns, and relationships; exploratory data analysis; word sequencing; detection of anomalies, irrelevant features, and data outliers; topic-based information collection; rule-based expert system associations; decision support systems; customer service chatbots

Dimensionality reduction for generalization of machine learning models; acceleration of gene expression analyses; segmentation of customer data; discovery of latent features within complex datasets

- *Reinforcement learning*

Reinforcement learning is a machine learning technique that uses rewards and punishments to train an *agent* (the doer of an action on behalf of another entity). In the context of AI, an agent is an autonomous and evolving system that perceives sensory inputs and interacts with its environment to achieve goals. These goals are encapsulated mathematically in the agent's *objective function*. An increased output value for the objective function signifies a network reward configuration, and is retained, or learned, by the agent. Agents are also trained to make decisions in accordance with rule-based *policies*, often ethical in nature, specifying what actions should be taken under different circumstances.[156]

Reinforcement Learning Applications:

Robotics, process optimization, gaming, chatbots, autonomous vehicles, natural language processing

- *Neuroevolution* – See Section 4

ACTIVATION FUNCTIONS:

Activation functions determine whether and how to involve downstream layers in a neural network's processing operations. Since these determinations are based on synaptic weights, the weights must be fine-tuned throughout the network to obtain the best results. This is accomplished by training the network to arrange and learn the optimum configuration of weighted neurons, typically involving backpropagation. However, linear transformations, such as the basic transfer function $\Sigma X_i W_{ij} + b$, can only support the training of trivially simple models through back-propagation. In contrast, activation functions $\varphi\ (\Sigma\ X_i\ W_{i\ j} + b)$, are nonlinear, and thereby possess the differentiability necessary to facilitate gradient descent and backpropagation through complex models.

- *Sigmoid Activation*

Inspired by the activation potential in biological neural networks, the sigmoid activation function introduces nonlinearity into feedforward neural networks which is essential for handling complexity, and the network's learning and predictive capabilities. The sigmoid function maps input values to an output value between 0 and 1.

Sigmoid Activation Function Applications:
Sentiment analysis, fraud detection, disease diagnosis

- *Tanh (hyperbolic tangent) Activation*

The tanh function has a steeper gradient than the sigmoid function, making it more useful for deep, recurrent, and long short-term memory neural network applications. It outputs values between -1 and 1.

Tanh Activation Function Applications:
Sentiment analysis, recommendation systems

- *Rectified Linear Unit Activation*

A piecewise linear function that outputs the input directly if it's positive, and zero if it's negative. It helps mitigate the vanishing gradient problem (During training, the network's weights are updated based on the gradient of the loss function. As the network's depth or sequence length increases, the gradient magnitude typically decreases, which can slow or stop the training process.). The rectified linear unit activation function improves network performance and accuracy in certain deep learning applications.

Rectified Linear Unit Activation Function Applications:
Image recognition, natural language processing

- *Softmax Activation*

An activation function typically used in the output layer when doing multi-class classification tasks. It provides a categorical probability distribution for each class in the output layer.

Softmax Activation Function Applications:
Image and speech recognition, natural language processing, probability-based predictions

NETWORK STRUCTURE

- *Feedforward Neural Network*

A type of artificial neural network where data moves in only one direction—forward—from the input layer, through the hidden layers (if any), and to the output layer. There are no cycles or loops in the network. Feedforward neural networks were the first type of artificial neural network invented.

Figure 12-4 depicts a *fully connected* feedforward network in which every neuron in a layer is connected to every neuron in the immediately previous and subsequent layers. In a *generalized feedforward network*, by contrast, data may be directly fed-forward to neurons in any subsequent layer. Figure 12-4 further illustrates that while the *architecture* of a feedforward network consists of unidirectional data flow from input to output, the *algorithm* of backpropagation, which fine-tunes the data content of network, operates in the opposite direction.

Feedforward Network Applications:

Classification, predictions, decisions, regression, pattern recognition, computer vision, natural language processing

- *Recurrent Neural Network*

Includes cyclic data pathways, or loops, *within* a layer. Loops equip recurrent networks with the ability to supplement feedforward processing with stored information that can support temporal behavior and understand sequences of events and data. This internal memory and feedback capacity makes recurrent networks ideal for applications involving time-series data, such as natural language models where selection of the next letters or words depends on the context of what occurred before.

In a recurrent neural network, the output of a hidden layer is applied back into that hidden layer so as to join it with the next feedforward input (Figure 12-5, illustrating the Elman recursive network algorithm).[157] The internal state of a layer is preserved by a set of context neurons which influences the result of subsequent inputs to the layer, thereby generating context-based decisions and actions.

A type of recurrent neural network called the *Long Short-Term Memory* network facilitates learning of long-term dependencies. These networks store and access long-term dependencies using a memory cell and three gates: an input gate, an output gate, and a forget gate.[158] The cell, configured much like conventional computer memory, stores data over arbitrarily long time intervals, while the gates regulate the flow of data on the basis of salience-weighted values.

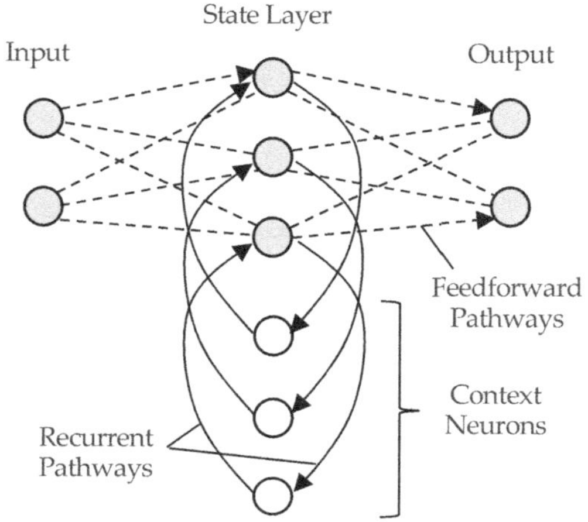

Figure 12-5 *Recurrent Neural Network*

Recurrent Network Applications:

Natural language processing and speech recognition; time series analysis and predictions, such as stock market fluctuations; robotics; video games; and healthcare

- *Convolutional Neural Network*

A convolutional neural network is a type of feedforward network that operates through a hierarchy of feature transformation layers to produce an output. These networks are especially useful for finding objects, classes, and categories in images and patterns that have a grid-like topology.[159] The basic principle is to train the network using a large dataset of labeled images, e.g., cats and dogs, so that it learns to identify and classify new images of cats and dogs. (The word "convolve" comes from the Latin *convolvere*, meaning to roll together, entwine. Mathematically, a convolution refers to how much two patterns—an image and a feature detector—correlate as one superimposes upon the other, which in AI applications is calculated by matrix multiplication.).

A convolutional neural network consists of the following components: a large input dataset, feature detectors (filters), feature maps (convolutional and pooling layers), a flattened layer, a fully-connected layer, and output results. The first layer of the network is a convolutional layer formed by applying a series of filters to an input image grid (image pixel configuration). Filters are smaller grids that pass over, or "stride" across, the input image grid, checking for the presence of mutually held feature patterns. In the early layers of the network, rudimentary features (edges, colors, gradient orientations) are identified, while in the deeper layers, higher-order features are progressively detected and merged (eyes, paws) to produce output results (cats, dogs).

Historically, in primitive networks, filter characteristics were put in by hand. But in modern networks, filter characteristics are obtained through supervised learning; i.e., the optimal weight values for filter matrices are learned through backpropagation.[160] Mathematically, a dot product (superimposition) is calculated between (the binary representations of) the image pixel matrix and a filter matrix, which is fed into an output array called a feature map. This process is known as a convolution. An activation function is applied to the convolved image to add nonlinearity and the capacity for complexity. The convolution process repeats itself until the filter has swept across the entire image, generating a stack of feature maps.

The feature maps are then dimensionally reduced—by reducing the number of features—onto pooling layers which retain only the essential features of the map required for image classification (or other objective), thus improving the model's computational efficiency. The pooling operation is similar to convolution, sweeping a filter across its entire input layer and using a rectilinear activation function to introduce nonlinearity. However, a pooling filter does not have any weights to assign, but instead applies an aggregation function to populate its output array. With continuing iterations of this convolution/pooling layering process, the network identifies more abstract portions of the image.

The highest-level stack of convolution/pooling layers is now flattened into a column vector (the minimal matrix representation), which is fully-connected to a standard feedforward network layer; this is where back-propagation commences for iteratively optimizing weights with respect to the input image and the filters,[161] and reflects the learning of the non-linear combinations of features represented by the outputs of the convolutional layers. Finally, the fully connected layer leverages the softmax function to generate probability values (confidence levels) for the various classification outputs learned in the convolutional layers (cat=1%, dog=99%).

Convolutional Network Applications:

Object and image recognition, classification, clustering, and segmentation, including medical imaging analysis and tumor detection; multidimensional data analysis; text classification, sequence labeling, sentiment analysis, document categorization, and speech recognition in natural language processing; brain-wave activity patterns; time-series data such as sensor readings (temperature sensors accelerometers, gyroscopes,), financial data, and physiological signals; anomaly detection; predictive maintenance

- *Transformer Neural Network*

Transformer neural networks are designed to *transform* input sequences of data into output sequences in applications such as natural language processing and computer vision. We will describe the structure and properties of the transformer in terms of natural language processing, where lexical strings (meaningful words, phrases, and sentences) represent the sequential data.

The main components of the transformer are an *encoder* to process the input sequence and a *decoder* to generate the output sequence (Figure 12-6).[162] After this algorithm has been coded into a programming language, it is trained on a relevant, usually massive, corpus of text to obtain the words that constitute the model's vocabulary. Each word is then converted into a scalar-valued number, or *token,* that identifies the word. We will use the "word2vec" method and adopt its word-based tokenization. In general, however, text sequences may also be tokenized into words, sub-words, punctuation, and white spaces in order to optimize computation efficiencies.

Once the vocabulary has been established and tokenized through training on a dataset, each word (token number) is *embedded* into a multi-dimensional vector space, where the dimensions correspond to the semantic features of words. Within the embedding space, word-vectors are correlated with other word-vectors to capture their semantic inter-relationships and grammatical patterns[163] in a format the model can process. (Note that an artificial neural network processes only the syntactic relationships used to the structure meaningful semantics which can be understood by the human mind, exclusively; see the discussion on qualia in Section 4.)

Each dimensional component of an embedded vector carries a weight, or strength, with respect to a particular feature of the corresponding word. Initially, weights are assigned randomly, followed by updating through cycles of training with backpropagation refinement. On large scale applications, vectors are stored in vector databases, linking vast repositories of precomputed context-based embeddings relevant to the selected data environment (this process is also applied to image, audio, and other complex data types). On the internet, vector databases are used to structure the semantic web by binding disparate elements into a cohesive whole.

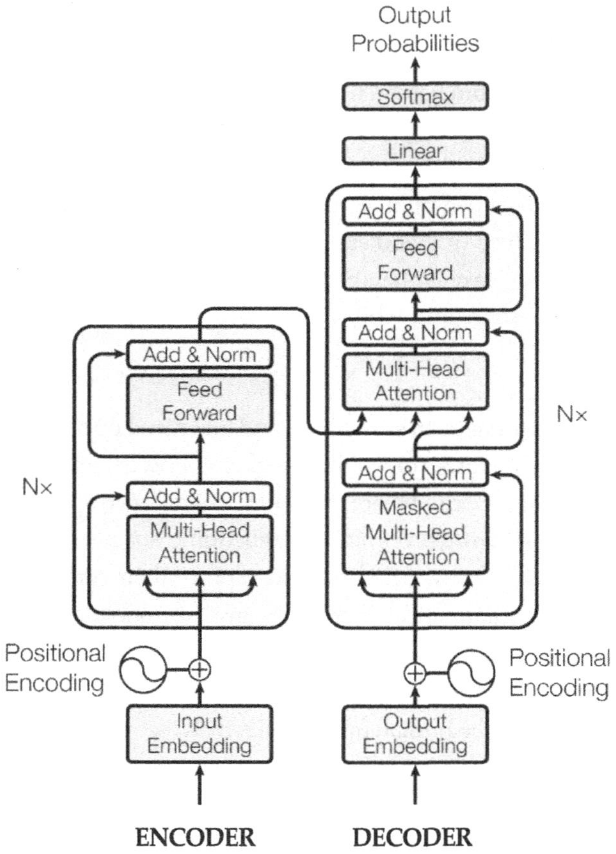

Figure 12-6 *Transformer Neural Network*

A primary mechanism of the transformer is *multi-head self-attention.* The term "attention" refers to a technique that enables the network to focus on the most relevant parts of a data sequence.[164] "Self" means that the network focuses only on the sequence of words currently being processed, exclusively addressing how distant words within that sequence influence and depend on one another. The term, "multi-head," refers to the tactic that all words in the sequence are processed simultaneously, thereby harnessing the efficiencies of parallel processing. Since the words in a sequence are processed in parallel, and therefore the significance of word order is not regarded, a positional encoding function is incorporated into the algorithm to retain the contextual importance of word proximity.

Transformer Architecture - The following description is based on the original transformer paper, *Attention Is All You Need,* which addresses an encoder-decoder transformer architecture. However, encoder-only and decoder-only architectures are better suited for certain tasks. Encoder–only models focus on applications such as language *comprehension* such as word classification and question answering, while decoder-decoder systems focus on content *creation* in usages such as chatbots.

In the encoder-decoder architecture shown in Figure 12-6, the attention mechanism is applied in three interrelated ways: encoder-encoder self-attention, decoder-decoder self-attention, and decoder-encoder cross-attention. In encoder-encoder self-attention, all words in the input sequence are processed and correlated simultaneously to produce an overall input context vector. In the decoder, words are generated for the output sequence one at a time, so that decoder-decoder self-attention concerns only the context of the words it has already generated, masking any possible influence of subsequent words. In decoder-encoder cross-attention, the decoder integrates the input context vector produced by the encoder into its processing stream to maintain and guide a contextually coherent associativity between input and output sequences.

The attention mechanism involves finding the most contextually relevant semantic relationships among the words that compose a textual sequence. The word-vectors embedded in a multidimensional space during training have both a semantic denotation and a geometric representation. They may be viewed as vectors in a geometrical space where semantically similar words have high weight components in the same feature dimensions, and therefore a high dot-product (degree of alignment) in these common dimensions. With high weight components and a high degree of alignment in the same dimensional sector of the vector space, also delimited by the context of the textual sequence, vectors that denote mutually relevant words are said to be "close" to each other in the space. Word-vectors that are close, i.e., relevant, to each other in this space should pay attention to each other. In the transformer, attention is brought about as the result of a mathematical formula operating on the word-vectors in the geometrical space.

The mathematical formula for attention is:

$$\textbf{Attention (Q, K, V) = softmax } (QK^T / \sqrt{d_k})\ V$$

--Q is the *query* vector. Q represents a particular word in a sequence for which the contextual relevance of each other word in the sequence is sought. Every word in the sequence queries every other word in parallel.
--K is the *key* vector. K is used to compute the relevance of a word with respect to Q. K indicates the location of relevant information in a vector database.
--V is the *value* vector. V is the actual value behind a key. The value is the meaningful data retrieved from a trained database once a query has selected the relevant key.
--QK^T is the dot-product of the vectors Q and K, representing the relevance of K to Q (the superscript, ^{T}, indicates the transpose of K, a required operation in matrix multiplication). QK^T computes the attention table, also called the attention score matrix.
--d_k is the dimension of the key and value vectors. This term is introduced to reduce the magnitude of attention scores and decrease the occurrence of vanishing backpropagation gradients.

--softmax is the activation function described above. It converts the table of results from $(QK^T / \sqrt{d_k})\ V$ into a probability distribution, from which the word denotation with the highest value is selected to be the next word in the output sequence.

Transformer Network Applications:

Natural language processing - text summarization; text generation: compose stories, essays, and poems; answer questions, translate between languages, chat with humans; natural language interface with agents (see reinforcement learning for description of agents);

Vector databases are useful for finding items conceptually similar to a given query, with applications such as semantic search, image search, and recommendation systems.

The transformer quickly captures interactions between different parts of any input, such as the components of a sentence, notes in a musical score, pixels in an image, or the atomic groupings of a protein or DNA molecule.

* * *

3- Biological Neural Networks and Mental Consciousness – Ultrasensory Theory

The features of ultrasensory theory described in previous chapters most relevant to the attainment of consciousness by an artificial neural network are reiterated in this section for convenient reference. In Section 4, the implications of this review regarding artificial consciousness will be discussed, noting the similarities and differences between biological and artificial neural networks, and whether artificial systems can be programmed or evolved to attain the complexity necessary for holistic closure and the emergence of subjective mental experientiality.

Biological Ultrasensory Theory

After the chemically rich planet earth had consolidated under the force of gravity about 4.5 billion years ago, a tranquil environment hospitable to electrochemical interactions was established. Interaction pathways closed to form autonomous, self-reinforcing autocatalytic cycles competing for limited chemical resources. The fittest cycles survived and evolved into *holistically complex* organic cells (holistically complex systems are unified, self-proficient entities possessing emergent properties distinct from those of its interdependent parts). These living, unicellular organ-isms were able to self-encode and utilize knowledge acquired from their prior experiences, thereby expressing the proficiency of *elementary consciousness*. For example, an amoeba that can seek out, recognize, and engulf paramecia, while continuously improving its survival skills on the basis of accrued knowledge, displays elementary consciousness. Within colonies of unicellular organisms, the selective benefits of synergistic collaboration induced a second level of holistic closure in the form of multicellular organisms, which expanded the capacities of elementary consciousness. However, these organisms lacked the neuronal means to be mentally aware of their experiences, for such an aptitude would require yet another level of complexification and holism—and the realization of *mental consciousness*.

The holistic multicellular platform provided an environment where cells adapted for the selective benefits of information production, transfer, and integration. In several types of primitive organisms, sensory-motor cells divided into discrete sensory and motor cells as well as interconnecting *principal neurons*. This laid the foundation for the simple nerve nets of animals such as the hydra. In a subsequent stage of evolution, a subset of the sensory cells further bifurcated to produce *interneurons*, the foundation for the more complex nervous systems of animals.

Interneurons and short-range principal neurons form local microcircuits that communicate via long-range principal neurons. Microcircuits regulate the flow of signals through widespread logic and control systems, and coordinate the global synchronization of firing patterns. Specifically, microcircuits send out messages about local firing patterns to remote targets. These targets return messages reciprocatively, inducing the firing patterns of the local and distal structures to synchronize, or phase-lock, their oscillation frequencies, generating a spatial network of temporally organized, endogenously causal process-structures.

Interneurons are specialized to manage the integration of information, as well as to assemble complementary structures able to collectively sense, or ultrasense, that information. The collective sensation of integrated information is implemented exclusively within an emergent, novel realm of neuronal functionality, the interneuronal domain, where mental consciousness is made possible. (In vertebrae, the interneuronal domain is located entirely within the skeletal confines of the central nervous system, whereas the peripheral nervous system is absent of interneurons.) Moreover, the ultrasensory collective structure, assembled in concert with its complementary informational representation, is the entity that carries the highest form of experientiality, having the phenomenological essence that we realize in our waking and dreaming states of mental consciousness.

Interactions between collective structures and integrated information are fundamental to both consciously experienced and subliminal neural operations. For an informational representation to be consciously experienced, it must attain a certain threshold of stability. Experiments in cognitive psychology have correlated consciously reportable events with *sustained, reentrant* (recycling) states of informational representation; whereas brief or variable representations are always subliminal (although they may affect sensory-limbic, executive-motor, cognitive, memory, and other functionalities). Over extended durations, the reentrant cycling of information through a series of incrementally updated stable states has the effect of endowing mental consciousness with a coherent temporal flow.

The state of mental consciousness is produced through the process of *neuronal group selection* (also called *neural Darwinism*). The term "neuronal group" designates the minimal, indispensable neuronal structure able to support a particular, content specific, experiential state. To maximize synergistic benefits, the intricacy and characteristics of the experienced informational representation are regulated in conjunction with its complementary, experiencing, collective structure (together constituting the totality of the neuronal group), as well as nervous system performance and related functionalities.

The process of neuronal group selection has evolved a repertoire of strategies which includes: *resource allocation*— apportion limited neuronal resources to optimize overall (subliminal and experiential) efficiency and content; *uniqueness*—construct only one neuronal group at a time to ensure largescale consistency; *saliency*— experience only the most urgent, important, or relevant information; and *unitarity*—experience a coalesced, transcendent form of information enriched with quality and meaning. For example, when the sight of a person's face is perceived, its shape, color, and other qualia are always bound together and experienced as a whole. A unitary experience may encompass a wide variety of informational categories, qualities, semantic meanings, memories, and intentional impulses within a single representation. Hence, the sight of a face might induce recognition, related memories, feelings of emotion, an urge to act, and so forth, into a unitary experience.

The experiential repertoire encompasses integrated combinations of basic sensational qualia (the redness of red, the feeling of pain), emotions (anger, love), meaningful constructs of the intellect (the notion of justice), cognized logical flows through time (a mathematical computation, the thread of a conversation), imaginations, and, importantly, discernments of intentional projections from the first-person mental self toward the third-person noumenal world, both in the form of executive-motor actions directed outward to the external world and top-down projections directed within the world of mind and body. These, and all other varieties of subjective, first-person phenomena are denoted comprehensively by the term "qualia." Qualia are often taken to refer to only the most basic raw feelings, which may then act as semiotic signs that trigger cognitive processes which produce meaningful cognitive states. We shall adopt the former, comprehensive convention; see Section 4 for a detailed discussion of qualia.

Mental consciousness has three experiential components: the *experiencer* (the collective ultrasensory structure), the *experienced* (integrated information), and *phenomenal experience* (the essence of mental consciousness that supervenes upon the experiencer). The interface between the experiencer and the experienced consists of the group of dendritic arbor synapses that are active during an event of mental consciousness. Additionally, dendritic arbors are the repositories for memories encoded in the configuration of synaptic spines that stud the surface of each dendrite.

The collective structure of the experiencer consists of a patterned array of post-synaptic dendrites, and the informational representation it experiences is rendered by an inversely patterned array of pre-synaptic axons. Signaling between the experiencer and the experienced is conducted both through and around the core synaptic interface. Feedforward signals transmitted from the pre-synaptic side of the interface bypass it in order to shape the collective structure directly, priming it to interact with only the most salient and contextually relevant informational signals passing through the interface. In turn, feedback signals transmitted from the postsynaptic side of the interface cycle back and reenter the pre-synaptic input stream, prioritizing, filtering, and qualifying the very (sustained, recycling) information being experienced by the collective structure.

In a highly evolved neuronal group, the signaling pathways traversing the core synaptic interface are massively parallel, bidirectional, and profusely interconnected, establishing an architecture based on coupled and nested cycles. At this level of complexity, the physical distinction between the experiencer and the experienced is inscrutable. Nevertheless, while the core interface may be subsumed in the complexity of the neuronal group, the reciprocative duality between the experiencer and the experienced retains its archetypal, causative influence on the third component of mental consciousness—phenomenal experience—which has its own subjective ontological nature.

Mental consciousness reflects two ontologies:
1) The third-person objective ontology, which refers to indirectly surmised "knowledge" about the content of the reentrant informational and collective structures (the physical "reality"), derived from empirical observations and inferences that have been correlated with experiential events.
2) The first-person subjective ontology, which refers to information as it is directly known by the mind in the form of qualia (the experiential, phenomenal reality) stemming from the endogenously projected perspec-tive of the collective structure on the integrated informational structure.

Longstanding attempts to explain experientiality in terms of the third-person ontology alone have failed to solve the "mind-body problem," i.e., how to bridge, via a continuous causal chain, the "explanatory gap" that separates the physical brain from the phenomenal mind. This gap is illusory, however; dispelled by recognition that both the first- and third-person ontologies refer to the same physical state. The one physical state *is* the bridge that connects two perspectives. The bridge does not connect the physical brain to a distinct experiential mind, rather, it connects the third-person, objective ontology conjectured about a physical state, to the first-person subjective ontology based on that state's experiential essence.

4- Artificial Consciousness

A variety of theories on the subject of human consciousness have been proposed, including: Orchestrated Objective Reduction Theory,[165] Global Workspace Theory,[166] Global Neuronal Workspace Hypothesis,[167] Integrated Information Theory,[168] Integrated World Modeling Theory,[169] Computational Theory of Mind,[170] and Dendritic Integration Theory.[171] These theories support facets of the ultrasensory theory of consciousness, outlined in Section 3 and detailed throughout this book, but they do not specifically consider how mental consciousness emerged, nor do they provide a satisfactory description of the essence—the experiential reality—of mental consciousness. The ultrasensory theory of consciousness identifies both the origin of consciousness (the emergence of the mentally consciousness brain at the holistic closure of the interneuronal domain), as well as its essential nature (the property of ultrasensation that transcends basic neuronal sensation).

For an artificial neural network to be conscious in a manner similar to the brain suggests that it possess certain characteristics of the brain's neural network which are essential to its consciousness. Our goal is to specify some critical characteristics, identify the ones currently lacking or undeveloped in artificial networks, and consider how they could be incorporated into an artificial network so as to promote the emergence and evolution of consciousness.

Characteristics of the brain's neural network currently being developed and implemented in artificial neural networks include:

1- Neuroevolution
2- Complexification
3- Unified processing and memory
4- Neuromorphic computing

Characteristics of the brain's neural network that do not currently exist in artificial neural networks include:

5- Collective ultrasensory structure
6- Holistic foundation
7- Self-referentiality and meaning
8- Qualia

1- Neuroevolution

Neuroevolution[172] is a machine learning technique that simulates biological natural selection and reproduction. A neuroevolving artificial network may incorporate genotypes, phenotypes, selection, adaptability, fitness, mutation, crossover, and novelty. A neuroevolving network starts with the specification of a genotype, which is the encoded representation of the network's architecture and parameters. The genotype is then repeatedly modified by genetic operators, such as mutation (random alterations) and crossover (swapping elements between genotypes), to produce a population of outputs, or phenotypes.[173] The fitness of each phenotype is evaluated based on the quality of its performance, and the genotype that produced the fittest phenotype is selected to propagate subsequent generations of networks.

Neuroevolution has some advantages over other machine learning techniques.[174] In traditional weight-based training methods, gradient descent is generally used to find a local optimum, which is not necessarily the global optimum. In neuroevolution, the maximum fitness for a given task involves a search through a population of network outputs, effectuating a global search capability. Neuroevolution algorithms thereby find the set of weights that approaches the global optimum, as well as circumventing the difficulties of exploding and vanishing gradients which arise when optimizing an individual network.

Neuroevolution algorithms simulate both the cooperative and competitive bioevolution selection paradigms.[175] For example, "swarm intelligence" mimics the cooperative behavior of flocks of birds and colonies of ants[176] by means of a population-based algorithm called "particle swarm optimization" (of individuals or agents). On the other hand, competition-based algorithms employ the strategies of reward/penalty reinforcement learning and generative adversarial networks.

A group of neuroevolution algorithms have been designed to discover novelty, rather than optimize fitness. Finding the fittest network for a given learning method limits its evolution to the *objective optimization* of the kind of task for which it was specifically intended. *Novelty search algorithms,* in contrast, are based not on the fitness of parent networks, but in rewarding their changeability. This strategy provides the potential for discovery, diversity, innovation, and adaptability. Novelty search has demonstrated that natural evolution's ability for diversification can be replicated, finding broader computational optima that go beyond the borders of "survival of the fittest."

In a technique that combines the advantages of objective optimization and novelty search, *quality diversity*[177] algorithms optimize fitness within a niche while simultaneously fostering diversification. The objective is to find a maximally diverse collection of individuals in which each member is high performing. Quality diversity algorithms, such as novelty search combined with local competition, balance a drive toward increasing diversity with localized searches for quality. In this way, evolution can move toward the discovery and realization of novel potentialities while retaining optimal fitnesses already discovered.

2- Complexification

A class of genetic algorithms have been designed to introduce complexity into artificial neural networks. Complexity is a crucial factor with respect to consciousness—the conscious human brain is the most complex system that we have yet discovered in the universe. NEAT (NeuroEvolution of Augmenting Topologies) is a genetic algorithm that incrementally adds new genes into a genotype for the purpose of increasing complexification[178] in the resultant phenotype (in neuroevolution, the term "gene" denotes a network parameter). NEAT has shown the ability to evolve increasingly deep and complex networks over generations.

Another method of neuroevolution designed to increase complexification entails the addition of feedback and recurrent pathways into artificial networks. In the brain, more than 60% of neuronal pathways involve feedback connections that have small synaptic weights.[179] These numerous feedback connections modulate already connected feedforward pathways. *Transfer entropy* is a measure of the interdependencies among neuronal pathways in a network. In artificial networks, transfer entropy is used to identify interdependent feedforward pathways, which are elaborated with randomly weighted feedback pathway connections. A genetic algorithm is then applied to obtain feedback connections with small weights, similar to the brain. This kind of complexification through feedback enhancement has produced significant performance improvements in different types of networks.

The brain has increased dramatically in size and complexity since the time of our distant ancestors. The cranium of the human brain now contains more than 100 billion neurons, interconnected by 100 trillion synapses, implying that its topology must have adapted through the course of evolution to accommodate these expansions ("topology" refers to the physical or logical configuration of a network, "architecture" refers to its theoretical framework, while "structure" encompasses both). Analogously, the topology of evolving artificial neural networks must be able to grow in order to accommodate increasing complexity. TWEANNs (Topology and Weight Evolving Artificial Neural Networks)[180] are genetic algorithms by which new neurons and connections are incrementally added to parent networks over generations in order to propagate off-spring networks that are progressively (physically and logically) larger and more complex.

3- *Unified Processing and Memory*

Conventional artificial networks and the brain differ with respect to information processing, memory storage, and their interrelationship. The basic computational element of a conventional computer system is binary, represented logically by 0 or 1, or true or false, signal gate open or closed; that is, whether a transistor is on or off. In contrast, the basic processing element of the brain's neural network is adaptable, the organic confluence of cyclic synchronization, excitatory and inhibitory neuro-transmitter admixtures, scale-invariance, and so forth. Further, each neuron in the brain—as well as every other cell in the holistic organism—has a unique DNA coding specialized to its function and location such that efficiencies gained through evolution are realized.

Conventional digital computer systems are also topologically inefficient: data is transferred between separate processing and memory locations via a single bus (pathway), resulting in the "von Neumann bottleneck," which becomes more severe as the size of external (memory, dataset, vector database) repositories increases. While there are forms of data and memory storage within artificial networks (such as recycling through context neurons in recursive networks, and long short-term memory), these are buffers which are transiently allocated data from *external sources* to suit the task at hand. In comparison, the brain's neural network utilizes bulbous structures, called dendritic spines, to store memory data at a micro-scale *within and throughout* its processing structure. Dendritic spines are located at the tips and on the shafts of a neuron's dendrites where signals are received from the axons of other neurons; in particular, in the highly complex reentry process transpiring across the core synaptic interface, as described in Section 3. A single pyramidal neuron located in the cortex or hippocampus—primary centers of associative memory function—can have thousands of spines arranged in very sensitive and plastic patterns. Thus, processing and memory storage are unified in the brain's neuronal substratum, while artificial neural networks based on the von Neumann computing architecture process data transferred from external storage sources.

The *memristor*, or memory resistor, is an electrical component that offers the potential to unify processing and memory within the substratum of artificial neural networks. The memristor is considered the fourth fundamental passive circuit element of electronics (in addition to the resistor, capacitor, and inductor). Originally described and named by Leon Chua in 1971,[181] it is characterized by its ability to change resistance based on the history of previously applied currents and voltages. It can be used as a switching element in artificial neural networks that, unlike the transistor, can simultaneously process and store data, retain its non-volatile state through a power supply interruption, hold more values than 0 or 1 (analog computing), facilitate massively parallel operations,[182] and be comparable to the neuronal synapse of the mammalian brain in size, energy efficiency, and processing throughput.

4- Neuromorphic Computing

Neuromorphic computing refers to hardware and software designs inspired by the brain's structure and operation. Processing and memory are collocated on neuromorphic chips (transistor- and/or memristor-based), eluding the von Neumann performance bottleneck as well as reducing energy consumption. Neurons are able to perform many functions in parallel, with all neurons and synapses potentially operating simultaneously.[183] Neuromorphic computers are highly fault tolerant; like the human brain, information is distributed over multiple locations so that the computer does not lose functionality upon the failure of some components.

Spatiotemporal organization and coordination in the brain, ranging from reentrant cycling through the core synaptic interface to frequency-specific brainwave functionalities and global harmonization, is essential for its consciousness-related operations. The patterns of integrated information that actually become the objects of consciousness experienced by an interpretant (experiencer and subjective meaning) are determined through the spatiotemporally coordinated process of neuronal group selection (Chapter 7). *Spiking neural networks* are artificial networks that incorporate temporal effects into their neuromorphic operations.[184] A neuron in (an analog implementation[185] of) a spiking neural network does not necessarily transmit information at each forward and back propagation cycle, rather it transmits information only upon attaining a certain "membrane" potential (a function that simulates the brain's voltage-based membrane signal conduction mechanism) by firing and conveying a signal to the membranes of other neurons, which, in turn, increase or decrease their firing potentials. Information encodement based on firing rate frequencies, spike intervals, spike train synchronicities, and other temporal factors, establishes a spatiotemporal platform analogous to the brain. Since only those neurons reaching their voltage threshold fire in response to spikes from other neurons, only a fraction of the neurons expend energy, making the network extremely efficient for the use of power. Other capabilities and benefits of spiking neural networks include: coherence among neuronal groups,[186] global network optimization,[187] combined encoding and processing of information, and representation of higher level, more abstract, symbols and functionalities.

In neuromorphic computing, electrical components analogous to inter-neurons, which form local microcircuits capable of coordinating the spatiotemporal global firing patterns that integrate information and associate widespread memory structures, are necessary to reproduce the underlying, unifying foundation of the brain. The memristor, together with the memcapacitor and meminductor,[188] provide a granular-level platform for the evolution of integrative artificial networks.

Object Orientation - The brain employs *object oriented* algorithms (Chapter 11). In contemporary artificial neural networks, the backpropagation (top-down) feedback process returns information from a final output result which progressively increases the intelligence of, i.e., trains, the primary (bottom-up) feedforward pathway. In the brain, the top-down executive-motor and intermediate pathways originate from endogenous sources, although they are inextricably intertwined with bottom-up sensory-limbic and lateral pathways. To mimic the brain, object oriented algorithms provide the ability to pervade the artificial network with function-specific processing modules at all levels. Object classes for network, neuron, and synapse,[189] messaging, encapsulation, inheritance, and polymorphism, elaborate the underlying straight-line (input to output) architecture of traditional artificial neural networks.

5- Collective Ultrasensory Structure

For an artificial neural network to acquire a form of consciousness similar to the brain requires it to have a collective structure dual to the integrated information it produces. As mentioned above, Dendritic Integration

Theory and Integrated World Modeling Theory (basically a combination of Integrated Information Theory and Global Workspace Theory) do not consider the presence of a dedicated receiving (afferent, subjective) collective structure for the brain, let alone for artificial neural networks. Rather, they focus on the production and processing of integrated information. Global Workspace Theory, for instance, proffers a "spotlight" that shines *onto* (not emanating from) selected representations of integrated information, making it available for cognitive processes—somehow transmuting that illumination of information into subjective experience. Similarly, Dendritic Integration Theory describes the thalamocortical system and the complexity of biological organization as underlying the production of a conscious state, but again, does not provide an intuitive explanation for the inward-coming nature of subjective awareness. Although a theory describing "categorizing sensory states" [190] does address the origin and evolution of first-person *global sensation*—which coevolved in conjunction with associative learning—it does not identify the specific neural network architecture and mechanisms by which the qualia of objects of consciousness are experienced.

The ultrasensory theory of consciousness delineates a collective structure that includes reentrant cycling through the spatiotemporally distributed core synaptic interface into associated afferent dendritic arbors. The collective structure possesses the supervening property of experientiality which emerged at the time of the holistic closure of the interneuronal domain. This theory provides an intuitive explanation for the inward-coming nature of ultra-*sensory* subjective awareness, applying to qualia related both to externally initiated bottom-up and endogenously created top-down information. Thus, the brain encompasses two complementary sectors. One sector generates integrated information while the other sector intakes *selected portions* of that integrated information.

The strategy is to design an artificial neural network initialized with ultrasensory characteristics in their embryonic form, such that those characteristics can develop as the network complexifies to holistic closure with emergent novel properties, possibly including consciousness. Any techniques discussed above for simulating the brain's evolution and operations, such as algorithms for complexification, spatiotemporal organization, unified processing/memory substratum, associative learning, etc., may be incorporated into the experiment as deemed efficacious. The primary architectural feature of this network is that it comprises mutually catalytic integrative and collective process structures evolving in tandem (Figure 12-7).

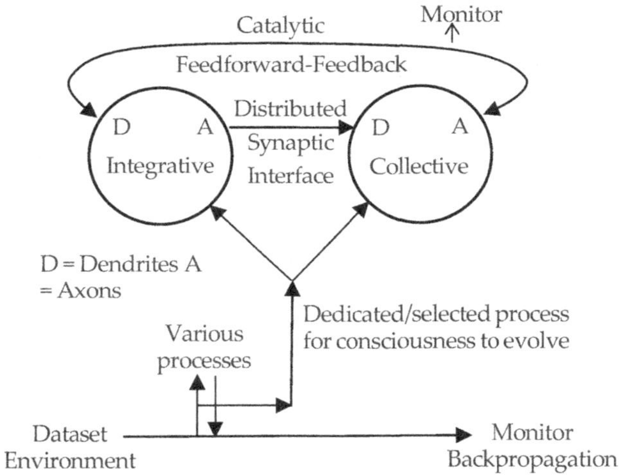

Figure 12-7 *Ultrasensory Neural Network*

In a neuroevolutionary artificial network initially partitioned into two sectors, with recycling through a core synaptic interface similar to the brain, the informational sector is intended to evolve into the *experienced*, and the dual collective sector into the *experiencer*. In both the brain and the artificial network, these sectors are physical/logical entities, the products of neurons and their interactions. *Experience*—the subjective awareness of qualia—is a non-physical (phenomenal) property that supervenes on the physical nexus of the experienced and experiencer. These three components do not initially exist in the artificial network, but are intended to emerge together in nascent form at holistic closure. Subsequently, the trait of experientiality can be sequestered into a partition of the integrative-collective structure for the selective advantage of unitary consciousness, as occurred for the biological brain (Chapter 7).

We conclude this section recognizing that it is unrealistic to organize a computer system with all of the characteristics listed above for the purpose of inducing its evolution into a holistic system comparable to the human brain. Already, it is not feasible to track the internal self-adjustments of a deep neural network fine-tuned by unmanaged techniques such as backpropagation[191]—described as a "black box" containing billions of parameters and nonlinear interactions.[192] However, another option for building an artificial network that could possibly evolve to become holistic and give rise to the emergent property of consciousness is to build a rudimentary biologically-inspired neural network modeled after the conscious organism Caenorhabditis elegans, which possesses the simplest known "true" brain. Of course, the result of this approach, like any other, is unknowable since it is not possible to confirm the subjective consciousness of another entity, even another human being. But it is informative to compare this possible evolutionary scenario for a proposed artificial neural network with the actualized evolution of the conscious brain.

The connectome of the nematode Caenorhabditis elegans has been fully mapped, revealing 302 neurons and 6702 synapses, making it the simplest extant nervous system with a brain. Its brain, with 188 neurons, is the central hub for sensorimotor functions, associative learning, short- and long-term memory, decision making, inter-animal communication, social behavior, and mating.

The nervous system of C. elegans is comprised of a nerve ring in the head region connected to a nerve chord that extends to the tail, together forming the central nervous system, where interneurons are exclusively located. The central nervous system, or interneuronal domain, projects to peripheral neurons and ganglia throughout the body. There are approximately 100 different types of neurons, including a variety of interneurons, which communicate through electrical synapses, neuropeptides, and a variety of excitatory/inhibitory neurotransmitters.

Neurons in C. elegans form local clusters of tri-neuron microcircuits consisting of various combinations of sensory, inter-, and motor neurons, which regulate long-range signaling, synchronize firing patterns, and integrate information spatiotemporally. In particular, two tri-neuron combinations are relevant to ultrasensation: an interneuron is presynaptic with two (executive) motor neurons, and an interneuron is postsynaptic with two sensory neurons; laying the foundation for an integrated information producing structure and an information receiving collective structure.

Neurons in C. elegans have dendritic spines able to encode primitive memories. However, there are no elaborate dendritic tree-like structures that can encode multifaceted memories as there are for the phylogenically older, but more complex invertebrate[193] planaria—dating from 500 million years ago[194] for planaria which possesses the oldest brain in the animal kingdom, versus 20 million years ago for C. elegans. Unlike Planaria, which has a cortex with dendritic arbors that refine and sharpen neuronal firing pattern representations (resulting from the demands of its free-living lifestyle, regeneration system, and complex body plan), C. elegans does not. Hence, for C. elegans, the interface between representations of integrated information (the experienced) and a receiving collective dendritic structure (the experiencer), is primitive and not fully formed. This implies that qualia (phenomenal contents of experientiality supervening on the physical network) are unsophisticated.[195]

The complexity of C. elegans' neural network is irreducible,[196] indicating holism and emergent properties, including and explaining the existence of the organism's (inferred) mental consciousness. The objects of consciousness produced by C elegans are rudimentary compared to those of the human brain which has evolved advanced competencies such as extended temporal duration of reentry cycles, multifaceted associations, and conceptual closure of mental representation. But despite these higher characteristics lacking in C. elegans, the most fundamental attributes essential for the emergence and evolution of consciousness, namely: elaborate catalyzation mechanisms, self-organized criticality, scale-invariance, associative learning based on dendritic spine dynamics, and holism, *are* found in C. elegans.

* * *

We now discuss characteristics 6 (holistic foundation), 7 (self-referentiality and meaning), and 8 (qualia), which are found in the brain but not in artificial neural networks, and present them as reasons why artificial neural networks based on conventional logic *cannot* evolve to achieve consciousness. These reasons are all related to holism: 6) artificial neural networks do not evolve upon and within a preexisting holistic infrastructure, 7) artificial neural networks that, today, are predominantly based on the Turing machine algorithmic paradigm precludes them from supporting the self-referential logic system of ultrasensory theory, and therefore artificial neural networks which deal in meaningless symbols cannot experience 8) qualia. A more detailed review and analysis of biological networks is now provided to more fully illustrate why duplicating their complexity, even of a single neuron cell, is formidable, and to explicate why conventional logic must be superseded by *cognitive logic* (described below) in artificial networks to acquire the crucial characteristics necessary for consciousness.

6- Holism

Catalytic processes have been instrumental in advancing simple open systems toward an attractor of complexity (an "open system" has variable contents as well as throughput from and to its environment). Upon reaching a certain threshold of complexity based on *cyclic* interactions among its parts, a phase transition occurred for many of these systems in nature, resulting in their holistic closure and the emergence properties and self-proficiencies distinct from those of its parts.

Evolution on the anthropic pathway has involved a hierarchy of holistic closures. Initially, molecules in early earth's environment formed cyclic interaction pathways that complexified until a phase transition engendered the emergence of chemical autocatalytic sets with self-proficiencies including self-organization and self-renewal. Building on this chemical autocatalytic foundation, the general principles of complexification and holistic closure rose through emerging levels of organization, where kindred holistic systems were collaboratively interconnecting for mutual selective benefits. In these cases, the interconnecting systems (parts) exported their individual holistic fitnesses to an inclusive, emergent, and transcendent holistic system (whole) with new properties and self-proficiencies. Specifically, holism ascended through the autocatalytic closure of chemical reaction networks, unicellular organisms, multicellular organisms, and the interneuronal domain; as described below:

Holistic closure of chemical reaction networks[197] - 4.6 billion years ago, catalytic metallic solids (e.g., iron, manganese, zinc, and their sulfides) were churning up from earth's molten core into the alkaline or lipid-based chambers that had formed in the carbon dioxide saturated ocean. These chambers established a tranquil environment for the formation of the reaction pathways needed for the advancement of complexity. A state of complexification was reached whereby each molecule in a cycle catalyzed the production of another component in the cycle, resulting in the holistic closure of a self-sustaining autocatalytic set having the proficiencies of self-organization and self-renewal (Chapter 4).

Holistic closure of unicellular organisms [198] - The origin of life involved a holistically closed autocatalytic chemical reaction network that attained the ability to both store information and self-replicate. Within this holistic chemical reaction network, a simple polymeric form of RNA arose which was both catalytic and able to encode information in its chain-like structure of nucleotide monomers. RNA diversified such that one variety, the ribosome, became proficient at catalyzing the production of proteins (phenotype), while the other variety (genotype) specialized to store and express information about the holistic interdependencies and functionalities of the set, including the genotype itself and phenotype production. The informational RNA evolved into the more stable double-stranded DNA structure (Chapter 4), capable of reliably encoding vast amounts of information—the DNA of C. elegans has ~100 million nucleotide base pairs, while the human genome has ~6 billion.

Holistic closure of multicellular organisms – Unicellular organisms with the same DNA genotype colonized for the selective benefits of synergistic collaboration, promoting inductive interaction (Chapter 4) among individual cells via the exchange of catalytic protein enzymes.[199] This led to complexification and a second level of holistic closure in the form of a multicellular organism, which subsumed the holism of the unicellular organisms (Section 2); each of the cells exported their holistic fitnesses to the multicellular organism, playing the role of holonomies in the newly integrated holism of the multicellular organism that displayed its own emergent self-proficiencies.

The intercellular relationships established during the complexification and holistic closure of the multicellular organism were incrementally coded into each constituent cell as a specialized variation of its DNA template according to its position in the multicellular body and corresponding functionality. In succeeding generations, the reproduction of the multicellular organism was, and is, initiated by the instructions encoded in the unspecialized DNA template of a single germ cell (lineage template), followed, for animals, by ontogenesis, in which the DNA of each replicating stem cell (template for specialization) is again altered according to its position and functionality, but now further modified by the "environmental" particulars encountered through embryonic and fetal development phases.

Holistic closure of the interneuronal domain – Is it possible that holism and consciousness could come to exist in information integrating AI machines through neuroevolution? [200] A wide range of theories for how an artificial neural network could attain holism has been proposed. A network could employ quantum effects such that the holistic superposed state enfolds decohering neuronal activities.[201] Quantum computing currently requires temperatures close to absolute zero, however, indicating that this approach will not be practicable until a series of major technological breakthroughs are realized. Another proposal is Holistic Graph Neural Network Theory,[202] which describes an attention mechanism based on a global context vector. However, graph neural networks mainly concern connections (graph edges) among local groups of neurons (graph nodes) as well as some global features of the graph, but do not take into account the scale-invariant interrelationships and properties that distinguish the holistic brain.[203]

The ultrasensory theory of consciousness provides a description of how the brain attained holism as a consequence of the successive autocatalytic closures of chemical reaction networks, multicellular organisms, and the interneuronal domain. In the interneuronal domain, the complexity of each prior stage of evolutionary closure was necessarily captured. In particular, the complexity of every DNA-based living cell, such as the ~85 billion neuron cells comprising the human brain, is essential to the support of mental consciousness.

The holism of the interneuronal domain arose through the complexification and closure of the *nodal network* (Chapters 6 & 7). The nodal network consists of localized process-structures, the nodes, which are interconnected through bidirectional pathways traversing and spanning both the cephalization hierarchy of the interneuronal domain, and the layered hierarchy of the cerebral cortex. This architecture is more precisely described as a system of cycles adjoining and unifying a constellation of convergence-divergence nodes; an open-system set of spatiotemporally coupled and nested cycles ranging from central pattern generators coordinating rhythmic motor activities, to interneuron-based microcircuits synchronizing global firing patterns, to reentry pathways cycling through the synaptic interface.

Catalytic processes in the nodal network involve *informational* representations and associations (contrasted to physical molecules and reactions in chemical networks). Neuronal catalysis operates at multiple levels of abstraction in the interneuronal domain. For example, neurotransmitters directly catalyze (activate and inhibit) relationships among neurons, nodes, and value systems; the informational food set is provided to the interneuronal domain (central nervous system) through signaling with the peripheral nervous system, which, in turn, connects to the extra-organismic environment. Biological natural selection involves a mutually catalytic relationship in which an organism both adapts to and molds its environment. For example, while the biosphere promotes adaptations in plants that optimize their homeostasis, the plants alter and regulate their biosphere through photosynthesis and transpiration.

Catalysis is also a central feature of the brain's cognition (information processing) system, both with respect to its evolutionary closure into a dedicated autocatalytic network, as well as to its dynamic operational processes. Mental representations (informational patterns representing concepts, ideas, memories, etc.) in the brain catalyze associations of other mental representations,[204] forming and learning new and more abstract mental representations. In turn, these representations reinforce the initial ones, creating a network of coupled and nested cycles[205] that has attained *conceptual closure.*[206] The brain continuously selects the set of catalytically associated cycles possessing the optimal information processing capabilities compatible with salience controls (such as the ascending value system illustrated in Figure 6-5, and teleonomic goals[207]) to maintain the state of self-organized criticality necessary for unitary consciousness. (A central property of a self-organized critical system is *spatiotemporal* scale-invariance, meaning that the system's structural and synchronicity motifs are uniform, or self-similar, through all process levels.[208] For the brain, self-organized criticality refers both to its underlying interneuronal infrastructure,[209] as well as to the instantiated episodes of consciousness that its mental representations support[210] (Chapters 3 and 6).

In artificial neural networks, activation functions perform the role of catalysts with respect to information processing. Activation functions currently used in machine learning, e.g., sigmoid, tanh, and ReLU, are inflexible compared to the many catalytic mechanisms employed by the brain, which range from neurotransmitters to abstract informational representations. Research and experimentation are underway to develop more diverse and adaptable activation functions for artificial networks to expand their abilities.[211]

For living systems, holism is inextricably embedded within the *physical* (spatiotemporal) aspect of their biotic architecture; [212] that is, it is derived from and dependent upon the complexity of life itself—the brain, in particular, has the most complex architecture of which we are aware.

However, this immense degree of complexity does not preclude the possibility that an artificial neural network could be devised on the basis of future innovations and discoveries that is able to attain physical holism.

As described below in 7) and 8), it is feasible to design an artificial neural network that could modify its input and embodied datasets continuously, via feedback from its optimized outputs, so as to coevolve with its environment. It is also possible for it to modify its own architecture to improve fitness and adaptability. It could increase the rate of emergence in general, promoting the production of novel properties and proficiencies. However, to attain *holism* similar to that of an organism such as C. elegans, all of the neurons in the artificial network would have to evolve holonomically, on the foundation of a common linking factor analogous to a DNA-based genotype that controls how each neuron develops in conjunction with all other neurons in the holistic phenotype. Moreover, to attain *consciousness,* the *logical* (informational) aspect of this abiotic architecture would also have to support the self-referential cognitive closure of meaningful information.

7- Self-referentiality and Meaning

The brain has evolved to process information on the platform of formal conventional logic. Conventional logic incorporates both inductive logic (draws the most probable conclusion from multiple facts or observations to form a general statement) and deductive logic (draws a specific conclusion by testing the validity of the general statement). AI machines also process information according to the principles of conventional logic in order to match the reasoning processes of, and the conclusions drawn by, the brain. However, conventional logic is incomplete in that formal systems cannot process self-referential information. The proof of Gödel's Incompleteness Theorem (Chapter 10), demonstrates that formal systems of logic introduce contradictions and paradoxes when dealing with self-referential[213] statements. In contrast, cognitive logic delineates the operations of the brain, a nodal network of profusely interconnected bottom-up *and* top-down, i.e., cyclic, neuronal pathways characterized by self-referential, holistically causal "axioms and inference rules."

Despite its incompleteness, the self-referential brain does *process* information on the basis of conventional logic to conform to the way events appear to unfold in the objective, inanimate physical world. Indeed, mathematical realism (the philosophy that the physical world operates according to mathematical rules which may be discovered by the mind) and mathematical idealism (mathematical rules are created by the mind and ascribed to the apparent behaviors of the objective world) specifically concern deductive logic, the foundation of mathematics in either case.

The brain itself, however, in its role as the system that implements conventional logic (what the brain does), has evolved to function according to the principles of *cognitive logic* (how the brain works). Thus, while traditional Turing-based AI machines function through algorithms structured on the basis of conventional logic (not including the neuroevolution pseudo-algorithms mentioned above, and which will be considered again below), the brain, and living systems in general, function through cognitive logic. That is: brain information processing, AI machine information processing, and AI machine functioning employ conventional logic, whereas the brain functions in accordance with the principles of cognitive logic. This distinction between conventional logic and cognitive logic with respect to the functioning of AI machines versus the brain has profound implications for the possibility of AI machines achieving consciousness.

Deductive logic (the rules by which the inanimate physical world works) and cognitive logic (the rules by which living systems work) are contextually different. Deductive logic is a deterministic system of reasoning that builds up conclusions from a set of axioms assumed to be true. Deductive logic is combinatorial, restricting algorithmic AI machines to a static foundation, only able to combine data and relationships already included, at least implicitly, in the original axiomatic ontology.[214] Similar limitations apply to inductive logic as well, which draws conclusions from the probabilities of already known facts and observations. Consequently, to endow an AI machine based on conventional logic with the ability to generate true novelty would require a set of axioms specifying every initial condition necessary to derive every unpredictable property, interaction, and relationship that could arise through the course of its evolution. Thus, AI machines cannot realistically discover or generate truly novel properties or relationships, as living systems working on the basis cognitive logic can.

In addition to their inability to handle self-referentiality, or to axiomatize evolution and novelty, conventional AI machines are, from the outset of their algorithmic design, not equipped to produce or entail subjective meaning. Machines deal with meaningless patterns, such as the pixelated patterns convoluted by image recognition algorithms, or the syntactic patterns of tokens manipulated in large language models. Since large language models can pass the Turing test, indicating human-like intelligence, we will compare them to biological neural networks.

Although the brain operates on the basis of cognitive logic, it employs the syntactic rules of deductive logic to process representations of meaningful concepts expressed by words, sequences of words, and other symbols.[215] It is only this syntax that is captured by artificial intelligence machines during training on a dataset. Once trained, the computer applies that same learned syntax to produce new sequences. Mathematically, the computer manipulates word-vectors that indirectly *denote* the semantic meanings of word sequences that can only be processed and *understood* through a system based on cognitive logic.

In contrast to the sequential, *straight-line* architecture of deductive logic, cognitive logic (a specialized case of systems-based logic[216]) entails the *cyclic* architecture inherent to evolving, adaptive, open systems, such as living systems and the brain—recall Morowitz's "fourth law of thermodynamics," stating that an open steady-state system (unchanging through a period of time) must contain cycles (Chapter 4). The cyclic architecture is able to represent integrated, scale-invariant, *object oriented* (Chapter 10) components and relationships. Organic cells, including neurons, specialize by means of the *object methods* encoded in their particular DNA gene expression. They perform functions collectively through *object encapsulation,* and exchange *messages* at many levels of encapsulation and group organization. Messages are *polymorphic,* causing various groups to execute different methods in response to a given message.

Messages transmit *meaning* relative to the methods encapsulated in the object groups of living systems. Moreover, the living system, as a whole, is an overarching group that respects the crucial meanings of life, such as self-referentiality, self-preservation, and self-organization—whether or not the system mentally understands those meanings. Approaches in AI programming, such as Deductive Object-Oriented Databases[217] and Frame Logic systems,[218] combine object structures with logical rules to model biological features like encapsulation, self-repair (methods), messaging, and so forth, but, again, they are processors of syntactical information with no *innate* semantic basis; conventional AI systems are disembodied and not rooted in lived meaning.

The holistic closure of living systems was accompanied by the emergence of cognition, the ability to process information about the environment, the self, and their interrelationships. All living systems are cognitive. For example, unicellular organisms possess the self-proficiency of *elementary cognition,* also called "basal cognition."[219] A unicellular amoeba can seek out and sense specific properties of a paramecium, process, i.e. cognize, information representing those properties, and choose to consume the food source for the self-referential purpose of autopoiesis. The radicles on the roots of a growing tree provide it with information it is able to cognize, empowering it to avoid nearby toxins for the self-referential purpose of survival. A human being can invoke its self-proficiency of *mental cognition* to process information about features on a map for the purpose of reaching a destination. A human being can also cognize information in the brain subliminally,[220] as when the word "thirsty" is quickly flashed on a movie screen prompting viewers to purchases drinks. In all of these cases, the cognized information is subjectively *meaningful* relative to the organism, whether or not it is, or can be, mentally aware of that meaning.

Beginning with the holistic closure of autocatalytic sets and the emergence of life, the evolutionary pathway for every adapting lifeform has been contingent on coevolution with and within its unique *umwelt*,[221] the portion of the environment it can perceive and with which it interacts. That is because lifeforms must be permanently embedded in their umwelt to maintain the low-entropy state of non-equilibrium necessary for self-organization and self-regulation (Chapter 14). This interrelationship affords a lifeform with the ability to leverage opportunities provided by its umwelt, enabling it to create novel, subjectively meaningful informational representations within its open-system ontology of scale-invariant coupled and nested cycles. This is not the case for artificial intelligence machines functioning on the static foundation of conventional logic. Network processes based on conventional logic are limited to first-order cybernetic feedback loops operating on non-embedded objective datasets, as opposed to the second-order cybernetic processes of systems which include themselves within the dataset—such as living systems—whose foundation of coupled and nested cycles enables both integrated objective (observed) and subjective (observer) processing.[222]

Semiosis is the process of creating and interpreting meaning through signs. Semiosis involves a triadic relationship between an object, a sign, and an interpretant (Chapter 7).[223] The object is a thing, independent of observation, such as a physical object, an event (e.g., an animal walks in the snow), or a mental concept. The sign, or representamen, is something that represents something else (footprints in the snow), such as a written word, a sound, or another sign. The interpretant is the meaning relative to an observer (interpreter) (infer the presence of an animal), which is variable depending on the constitution, beliefs, and cultural influences of the observer.

While semiotics is applied to both first- and second-order cybernetic systems, we will focus on the second-order cybernetic system of the brain, whose architecture supports a triadic relationship between integrated information (object), messaging (signs), and the meaningful subjective interpretation of those messages (interpretant). This will provide a clearer perspective on the limitations of first-order AI systems, and why paradigm reformation is necessary for them to evolve to attain a second-order subjective, understanding of meaning.

The three core components of semiosis, object, sign, and interpretant, are central to the ultrasensory theory of consciousness. *Ultrasensory logic*, a form of cognitive logic, is physically implemented on the semiotic neuronal platform of the brain. Ultrasensory logic is the basis for the brain's second-order cybernetic cognitive processing system, in which the observed is comprised of integrated information and the observer by the collective structure. For the human brain, the observed is actually the peripheral nervous system, as the central nervous system (interneuronal domain) has no direct contact with the external world. The semiotic process here is not only about creating an inner representation of the external world via the peripheral nervous system, but also about the continuous process of active inference dedicated to achieving the self-referential goals of the observer.[224]

In the case of the informational representations the brain cognizes for ultrasensation, the semiotic object is a sustained pattern of axonal firings, the sign is the assortment of neurotransmitters, released from the axons, flowing across and recycling through the spatiotemporally distributed synaptic interface, and the interpretant is the subjective meaning derived from the reaction of the neurotransmitters and the collective structure's dendritic ensemble; all transpiring in scale-invariant (fractal, Chapter 6) messaging dimensionalities. (Note again, this description is heuristic, referring to a reentrant process of coupled and nested cycles embedded within a self-referential open system that includes holistic causal forces; but the fundamental principle of sending and receiving messages across neuronal synapses still prevails as an export of fitness to the emergent ultrasensory process.)

8- Qualia

Mental consciousness is the subjective awareness of meaningful information produced by the brain. It transcends the objective meaningfulness that pertains to an organism's relationship with the environment or its internal state. For example, an amoeba can detect certain properties of a paramecium, cognize information representing those properties, and consume the food source for the self-referential purpose of survival, objectively meaning that it will survive. But the amoeba has no brain with which to mentally portray or understand that somatic meaning. In contrast, an organism with a brain can carry out similar objectively meaningful activities, *and* it can construct and portray information about those worldly events (as well as endogenously constructed information) in the form of mental representations which are subjectively experienceable by the *mind*.

The mind is the non-physical domain, or phenomenal reality, of mental consciousness that emerged with the holistic closure of the physical nervous system. It is characterized by a subjective qualitative ontology whose elements are meaningful qualia, the subjective "raw feels" of the informational representations produced by the brain. The "Mary's room" thought experiment[225] makes clear the nature of qualia: Mary is a brilliant scientist who has lived her whole life in a black and white room, never exposed to the individual colors of the spectrum. However, she has learned all of the physical facts ever discovered about the color red. One day she leaves the room and sees a red apple. She now knows, for the first time, "what it is like"[226] to experience the phenomenal quale of redness. She has learned, and felt, something new about the color red: that her first-person understanding of red has a qualitative *subjective meaning*, and that all objective physicalist knowledge of red is incomplete.

Mary's experience demonstrates that meaning and qualia are intimately related in the domain of mental consciousness.[227] Mental consciousness is not only about *what* an informational representation is to an experiencer (its meaning), but also *how* it makes the experiencer feel (the qualia). This transcendent mode of meaningfulness characterizes the subjective reality of a mentally conscious being. In the most primitive brains, qualia are unsophisticated, involving only basic feelings about exteroceptive stimuli (food chemicals, mating pheromones) and interoceptive bodily conditions (pain, hunger). These primal experiences of qualia thus refer to, transcend, and illuminate the meaningfulness of matters of elementary consciousness.

Qualia and meaning are not only associated with sensory information and other basic "raw feels." While the mechanisms and nature of mental content are a topic of vigorous debate in 2026, engaging such theories as "intentionality" (what a mental state is *about* or *of*), or "access consciousness" (concerning the availability and accessibility of information for conscious broadcasting), we use the term "qualia" broadly in order to capture the subjective connotations of *all* mental experiences. *Cognitive phenomenology*[228] is a philosophical specialism that advocates the comprehensive notion of qualia,[229] addressing the various types of information cognized by the human brain that are experienced as qualia. Cognitive phenomenology includes feelings that accompany: emotions (e.g., joy, fear); sudden insights ('aha' moments); intentions (purposefulness); and understanding (grasping an idea, an abstract thought,[230] or inter-societal observation or communication—the cultural environment of memes).[231] All of these examples of qualia signify the felt impressions that complement the cognized meanings of what we experience.[232]

* * *

The Road to Artificial Consciousness (and some speculations)

The characteristics of biological neural networks that made possible the emergence of consciousness may be stipulated as also being essential for the design of a conscious machine. Certainly, alternative design proposals are justified, but considering that the brain is the most complex structure of which we are aware in the universe, and that it is the only vehicle of mental consciousness known, it is naïve to expect that a hardware-software system with capabilities comparable to the brain could be devised ab initio with today's technology and programming expertise.

The eight essential characteristics that are necessary for a neural network to attain consciousness, as described above, can be grouped into two fundamental categories: holism and dualism.

Holism

In the biological domain, phenomenal reality (the mind) emerged with the holistic closure of the nervous system (the brain). Since it is not possible to predict the strong[233] emergent properties that appear at holistic closure, it is sensible to institute the same foundation upon which consciousness emerged in the biological domain in order to increase the probability that analogous properties would emerge for a machine attaining holistic closure. Thus, a *cyclic, catalytic, and object-oriented foundation* is essential, and neuroevolution techniques, such as: *complexification* (to self-organized criticality, NEAT, TWEANN), *Darwinism* (competition/cooperation, quality diversity), *processor-memory unification* (memristors and laminar cortex models simulating local networks of the brain[234]), and *neuromorphic computing* (spatiotemporal organization and spiking neural networks) would be invaluable for this project. Furthermore, these components and features must be interdependent and unified in a closed cause and effect cycle of scale-invariant relationships (closure to efficient causation[235]), as they were when the biological neural network attained holism. This requirement for an all-inclusive *relational foundation* is well beyond current machine design capabilities.

Dualism

Evolution to holistic closure is a necessary but not sufficient criterion for the emergence of consciousness. When a neural network attains holistic closure, certain dualistic relationships (having two complementary or opposing parts) must be in place so that properties such as qualia and mind can emerge and operate.

> *Science is built on dualities. Indeed, every mode of discrimination creates one. But the most fundamental dualism, which all others presuppose, is of course the one a discriminator makes between self and everything else* —Robert Rosen.[236]

In addition to Rosen's fundamental duality, five other dualities, which encompass the characteristics and features described above, are essential for mental consciousness; See Figure 12-8.

Before holistic closure

① Duality between the neural network and its external environment

② Duality between the efferent and afferent functions of neurons

After holistic closure and emergence

③ Duality between the integrated informational and collective structures (physical duality)
④ Duality between qualia and the mind (phenomenal duality)
⑤ Duality between the phenomenal and physical ontologies (mind-body relationship)

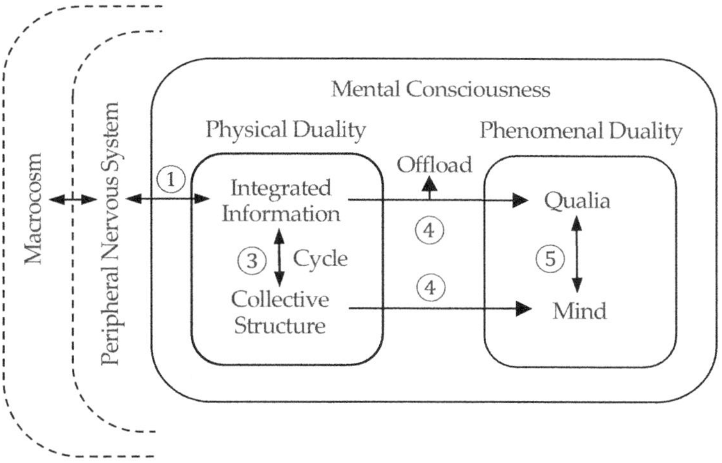

FIGURE 12-8 *Dualities Required for Mental Consciousness*

① *Duality between the neural network and its external environment*
Conscious living beings are open systems that maintain a reciprocative relationship with the external environment, or umwelt, in which they are inextricably embedded, and with which they must coevolve. The duality between microcosm and macrocosm (Chapter 3) is an essential factor for the emergence of life and consciousness in the biological domain. In the case of an open-system, evolving-to-consciousness artificial neural network (microcosm), its *dataset* represents the external environment (macrocosm) with which it must interface and coevolve in order to incrementally build up the complexity necessary for holistic closure and emergence.

The type of dataset that could most effectively coevolve with an open-system machine spans a wide range of possibilities. For example, interfacing dual generative models, with one network training, learning from, and complexifying the other, ensures synergistic evolution. A supplementary tactic would be to continue interfacing machines with human culture[237] (by means of robotic agents, artificial general intelligence devices, and direct communication with people). Human culture is already holistic, so this interface would remove the need to craft a coevolving environment for the open-system machine. The interface could be tuned for coherence at closure through the use of large concept models,[238] which, like people, process entire semantic concepts and ideas as semantic units.

In comparison to human culture, a network designed to interface with a limited umwelt dataset, say only visual images, would not be able to attain the complexity necessary for closure; nor would a dataset and interface designed to evolve a machine to a simpler vehicle of consciousness such as C. elegans, which has actually attained closure, would not take advantage of the monumental surge of efforts and advances presently underway (2026) in the field of artificial intelligence that are predominantly directed to human culture. It is likely that a combination of many tactics will provide the best opportunities to evolve a network-environment duality to closure.

② Duality *between the efferent and afferent functions of neurons*
This duality is a basic characteristic of biological and artificial neurons that manifests both internally and externally to the neuron. All neurons possess afferent dendrites whose threshold activation generates the firing of its own (dual) efferent axons. In turn, the firing of the neuron's axon influences the firing potential of other (dual) neurons to which it is synaptically connected. As a result of neuronal firings, information is transmitted and received throughout the neural network.

③ *Duality between the integrated informational and collective structures (physical duality)*

During the evolution of the nervous system, the selective advantages of improved information transfer via neurons drove their connectome toward complexity, as interconnecting feedback and feedforward pathways formed an elaborate network of coupled and nested cycles. When the connectome became sufficiently complex, the holistic brain emerged. The underlying duality principle ② of information transmission and reception across neuronal synapses was exported for fitness to the dual structures ③ that constitute the ultrasensory process (Figure 12-7). Specifically, messages from a highly integrated informational structure are transmitted to, and recycled through a dual collective receiving structure. For an artificial neural network, the emergence of the ultrasensory process could possibly be fostered by inaugurating a neuroevolution prototype that includes: the cyclic artificial neural network paradigm[239] (to provide an infrastructure upon which to build up scale-invariant feedback and feedforward circuits), classification algorithms[240] (to partition intermediate cycle nodes into efferent and afferent categories), and memristors (to embed memory into afferent nodes analogous to dendritic spines). Moreover, these and all other procedures and devices employed to replicate the brain's ultrasensory duality must complexify interdependently to facilitate the network's holistic closure.

④ *Duality between the physical and the phenomenal ontologies*

⑤ *Duality between qualia and the mind (phenomenal duality)*

Dualities ④ and ⑤ are discussed together because the phenomenal aspect of duality ④ is characterized by duality ⑤.

Upon holistic closure, physical duality ③ (between the integrated informational and collective structures) emerged in conjunction ④ with phenomenal duality ⑤ (between qualia and mind), respectively. "Qualia," described in Section 4-8, refers to the phenomenal content of a mental experience which supervenes on the informational structure, and "mind" refers to the phenomenal realm of the subjective experiences of qualia which supervenes on the collective structure.

The subjective, first-person perspective of the mind on qualia ⑤ is contingent on the underlying direct physical relationship ③ between the collective and informational structures that interoperate within the core of the open-system ultrasensory process. The collective structure has a distinctive identity with endogenous properties based on the functionally receptive dendritic aspect of the cyclic ultrasensory complex—where ultrasensed integrated information is shaped by recycling axonal output from the collective structure. Subjective experience involves the qualia that refer to reentrant information assembled on the basis of neuronal group selection, with other information offloaded and processed subliminally. The specific instances of information selected to be subjectively experienced as qualia by the human mind are sustained in a scale-invariant critical state for a duration of 200-400ms, the timeframe that has evolved to optimize holistic efficacy.

The evolutionary underpinnings that made possible the emergence of qualia and mind in living beings suggest a foundation upon which to evolve an artificial neural network to consciousness. Accordingly, holistic closure and the basic dualities are essential requirements. However, this, or any other, method for attaining mental consciousness—the pinnacle of evolution—in an artificial intelligence machine is certainly speculative and ambitious. It is offered here to emphasize the daunting challenges involved in the creation of consciousness, a feat that has taken nature billions of years of evolution on the anthropic pathway to achieve.

* * *

Part 3
Physics and Metaphysics

13

Gauge Theory

Gauge theory describes an intricate process first introduced implicitly in the ancient tattvic system, and now central to modern fields of science, including quantum theory, cosmology, and biology. Indeed, the success of gauge theory in relativity and the standard model has led to its prestigious claim as being the "cornerstone of modern physics." Gauge theory identifies animating stresses and informational linkages induced by a dichotomy inherent to the supersymmetry of the universe.

We begin our analysis with a naïve but heuristic glimpse into the origin of gauge symmetries by examining the relationship between global and local symmetries.[241] A *global symmetry* applies throughout an entire domain of applicability, or manifold, such as at every point in space or instant of time. A *local symmetry,* in contrast, is one that is free to vary from point to point across the manifold. The full supersymmetry of the universe encompasses both global and local symmetries, but the coexistence of global and local symmetries is unstable. This instability engenders the breaking of the global symmetry of the universe in the big bang event of a cosmic cycle, resulting in the gauge symmetry forces of nature that mediate interrelationships among the residual, unbroken, local symmetry spaces of particles.[242]

In the current state of the universe, the laws of physics are unfolding in a ten dimensional symmetry space that encompasses both global and local sectors. Four of the ten dimensions are extended, constituting the familiar spacetime that pervades the cosmos, while the other six dimensions are curled up within tiny "internal symmetry spaces." These local six-dimensional compact manifolds constitute the framework underlying the behavioral properties of elementary particles, such as electrons and quarks. The universe is understood to be ten-dimensional in the sense that a six-internal-dimensional particle also occupies a location in four-external-dimensional spacetime.

The *special theory of relativity* is based upon a global symmetry. This theory states that all observers moving through space, regardless of how fast they are moving relative to one another, measure the velocity of light to be invariant (3×10^8 meters/sec in vacuum, or "c"). Together with the symmetry principle stating that the mathematical form of the laws of physics must be the same throughout the universe, i.e., globally invariant, strange effects are predicted under this Lorentz symmetry. As an object traveling through space increases in velocity and approaches the speed of light relative to a given observer, its length appears to contract in the direction of its motion, while its mass appears to increase. Similarly, a clock in motion appears to run more slowly than one which is stationary relative to an observer's reference frame. Special relativity thus unifies features of nature that seem to be independent of each other in the low velocity world of our daily lives on earth, but space and time, and mass and energy, famously advertised by Einstein's formula, $E = mc^2$, are actually representations of a deeper symmetry principle.

The *general theory of relativity,* in contrast to special relativity, is based on a local symmetry. It generalizes the special theory to account for the presence of matter and energy in spacetime; in which case covariance (invariance with respect to the *form* of an equation expressing a law of nature) is dependent on a function of spacetime that varies from local event to local event (position in unified spacetime). The variable values of this "metric" function specify the local curvature of spacetime, which is proportional to the local strength of the gravitational force produced by the matter and energy in the region. Consequently, general relativity unifies curved spacetime continuum and the gravitational field.

The symmetry principles of special and general relativity, one global and one local, were both intuited by Albert Einstein. By assuming symmetry principles *a priori,* he subsequently derived the mathematical formulae that quantify and predict the dynamics of the relativistic universe. Since Einstein, conjectures about the symmetries of nature have led to the discovery of global isospin, the local electroweak interaction, and antimatter.

According to the theories of relativity, the laws of physics do not (and should not) depend upon the arbitrary choice of a *reference frame* [x] by an observer for the description of a physical system, be it a global or local reference frame. It arises, however, that a Lagrangian which has been constructed to express a given symmetry relative to a global reference frame cannot consistently describe the same dynamic system in a local frame (mathematically, the time derivative, i.e., the rate of change of the system's parameters, does not "pass through"). Therefore, historically, a compensatory term, called a *gauge potential,* was introduced, ad hoc, into the Lagrangian in order to preserve its form; that is, to permit the coexistence of nature's global and local symmetries within a unified physical theory.

In 1918, Herman Weyl investigated the gauge potential term in order to determine if it could include both electromagnetism and gravity in a single Lagrangian that was invariant under the transformations, or the "regauging," of spacetime *(eichinvarianz).* That idea proved to be non-viable, but in 1929, once quantum mechanics had been discovered, Weyl proposed, this time successfully, that the unifying gauge properties of electromagnetism are related to the wavefunction of matter particles.

---

x A *reference frame* is a system of coordinates that structures the independent dimensionalities of a symmetry space. An example of a curved-space reference frame is the system of latitude and longitude dimensions defined on the earth's surface.

The addition of a gauge potential term into a Lagrangian is more than a mathematical contrivance designed to accommodate the global and local symmetries observed in nature. The mathematical form of gauge potential terms also serves to codify the dynamics of the physical interactions that have resulted from cosmological symmetry breaking. This connection between the forces of nature and cosmological symmetry breaking is why gauge theory is so compelling. The principal value of gauge theory lies in its capacity to describe, predict, and integrate seemingly independent actions and features of the physical world. An example from particle physics demonstrates the underlying *physical* relevance of gauge potential terms: The mathematical inconsistency of the global and local (partial derivative) terms in the Lagrangian of an electron's wavefunction (the wavefunction describes the electron's location in spacetime) may be resolved by the addition of a compensatory gauge "vector potential" term which balances the Lagrangian symmetry. However, this purely quantum-theoretic vector potential term in the wavefunction, derived mathematically, is identical to the field equation of electromagnetism which had been deduced many years earlier by James Clerk Maxwell based strictly upon classical observations.

NOTE ON THE PHYSICS MODELS

The models described in *Consciousness* combine and compare features of well-regarded contemporary theories of physics, including relativity, quantum theory, the standard model, supersymmetry, dark energy and matter, M-Theory, and conformal cosmology. While some of these theories fill in conceptual gaps left by others, it may happen that their assumptions and basic ideas are incompatible; for example, quantum theory and general relativity in the microscopic regime. The actual details and refinements of the physics that will surely be discovered in the future are not crucial to the purport of this book. The physics presented here is intended to illustrate the broad range of ideas and models prevalent in modern theories, and to identify the underlying principles they share regarding to the nature and evolution of consciousness.

As of this writing, a number of theoretical conjectures in this discussion have not been empirically confirmed. In particular, signatures of supersymmetry and extra dimensions, which were mildly anticipated to be detected in the initial lower energy (8 TeV) runs of the Large Hadron Collider at CERN, have not been forthcoming; implying that these phenomena will emerge only at higher energies, or that other theories will be required.[243, 244] If supersymmetry and/or extra dimensions are not properties of nature, four-dimensional theories, such as conformal cosmology and Technicolor, will attract greater attention. Regardless of the existence of supersymmetry or extra dimensions, or the possibility of a multiverse capable of supporting many independent cosmic cycles, the fundamental symmetry principles relevant to the nature and evolution of consciousness as presented here—namely, symmetry Lagrangians, symmetry breaking, and gauge symmetries—would not be invalidated, as these principles are common to all of the above referenced theories.

* * *

In the current cosmological epoch, four symmetry groups are in effect: gravitational, strong nuclear, weak nuclear, and electromagnetic. Each of these symmetry groups contains interconnected global and local sectors. In the case of gravitation, the nature of the global and local sectors, as well as their interconnection, is unique, because gravitation is equivalent to the curved spacetime upon which the other three symmetry groups are defined. However, the non-gravitational symmetry groups do share a common framework, as formalized by the highly successful standard model. The standard model is a theory that encompasses two branches of physics: the microscopic standard model of particle physics and the macroscopic standard model of cosmology. Every high-energy physics experiment carried out since the mid-20th century has yielded results consistent with the predictions of the standard model, including the recent detection of the Higgs boson at CERN.

Unfortunately, the standard model leaves critical theoretical lacuna. It requires and implies, but does not account for, the origin of the Higgs field or the three generations of mass that have been observed for each species of particle. Also, conventional formalisms of the standard model are structured somewhat differently from gravitational relativity theory, an aesthetic deficiency for the ideal of an all-unifying symmetry principle. The *framed standard model,* however, takes a different perspective on the symmetries, proposing that the global and local sectors of each of the *four* symmetry groups are unified by means of a higher, interconnecting symmetry. This model thus takes a step toward the rapprochement of gravitation with the three gauge symmetries of the conventional standard model. The framed standard model further proposes that dynamical terms are contributed to the Lagrangian not only from the global and local symmetry sectors themselves, but also from the interactions between the underlying reference frames which parametrize these sectors. Thus, in the discussion that follows, we adopt the unifying physics of the framed standard model (unless otherwise specified), while also distinguishing the broken symmetry particularities of relativity from those of the standard model gauge symmetries.

In the case of gravitation, the magnitude of the gauge force is equal to the curvature density of spacetime; the greater the spacetime curvature, the greater the gravitational force. Spacetime curvature is generally specified by either of two related mathematical constructs: the metric or the connection. In fact, some theories, including quantum geometry, take the connection to be the ontologically fundamental element of spacetime. In simplified terms, the metric ("g") refers to the correction factor that must be applied to *lengths* measured by an observer in a curved spacetime using a locally flat, Euclidean reference frame, called a *vierbein* (Ger. four-legs). The connection ("∇") specifies the rate of change in the *orientation* of the local reference frame relative to flat planes tangent to the spacetime manifold. The mathematical relationship between the metric and the connection is $\nabla g = 0$. Both of these measures of curvature, i.e., the rate of deviation from flatness, correspond to the stress or strain[245] that gives rise to the gauge forces induced by the gravitational field. Understandably, in the early days of relativity theory, these ideas brought about conceptual struggles for even the greatest physicists, as illustrated by the following humorous exchange:

> *The whole thing is going through my head like a millwheel: To take the connection alone as the primitive variable or the metrics and the connections?*
>
> — Schrodinger, May 1st, 1946.

> *How well I understand your hesitating attitude! I must confess to you that inwardly I am not so certain ... We have squandered a lot of time on this thing, and the results look like a gift from devil's grandmother.*
>
> — Einstein, May 20th, 1946 [246]

The path an object actually traverses across the metric "surface" of space-time is the one that minimizes the change in the Lagrangian. This path delineates an extremum, a valley-like channel, called a geodesic, in the topology of spacetime through which the object effortlessly glides. Thus, the pressure felt in one's feet when standing on the earth' surface is actually an electromagnetic effect caused by the electrical repulsion between the molecules of the earth and those of the feet, each impeding the (gravitational geodesic) free-fall trajectory of the other. The realization that the action of gravity on a body is equivalent to the body falling freely, weightlessly, through curved spacetime, was said by Einstein to be his "happiest thought."

In the global sectors of the three standard model symmetries, specific invariant-valued properties are conferred upon classes of elementary particles (leptons and quarks) inhabiting the four-dimensional external frame of spacetime. These global properties are manifested as (1) the invariance of the number of particles in the universe (total baryon number minus lepton number), (2) the up and down "flavors" of particles, and (3) the three, and only three, mass generations of the elementary particles. The local sectors of the three standard model symmetries correspond, respectively, to (1) the electromagnetic, (2) the weak, and (3) the strong gauge symmetries. Together, the local sectors are confined to the internal, compact, six-dimensional reference frames, or *vielbeins* (Ger. many legs), of individual particles, whose phases elicit the gauge influences traveling the external space. Finally, the reference frame vector that unifies the global and local sectors also accounts for the presence and properties of the Higgs boson.

The behavior and strength of a particle in a gauge interaction depends on its charge. According to a theorem postulated by Emmy Noether, charges represent conserved quantities implied by the gauge symmetries (see Chapter 15). For example, the electrical charge of a particle connotes its participation in the electromagnetic interaction. Similarly, the isospin charge of a particle effects the weak interaction, while the color charge sources the strong interaction. For these internal symmetries, the gauge field influences traveling through the external space, mediating interactions, are called boson "particles," because these influences are always manifest in integer multiples of the fundamental quanta of energy and charge. The electromagnetic, weak, and strong bosons are termed photons, W and Z bosons, and gluons, respectively. The bosons are emitted and absorbed by another type of particle, the fermion particle (e.g., electrons, quarks, and neutrinos). We discuss the fundamental difference between the bosons and fermions below. The exchange of a boson between two fermions may either change the trajectories of the fermions, or transmute them into different fermionic species. For instance, the exchange of a Z boson between a neutron and a neutrino changes their trajectories, while the exchange of a W boson between a neutron and a neutrino changes them into a proton and an electron, respectively.

A particle is identified by the combination of the charge-types confined within its internal symmetry space. Numerous combinations of charge exist in nature, resulting in the menagerie of transient and stable particle species, dubbed the "particle zoo." These combinations lead to many, and sometimes counterintuitive, consequences. Some examples: A neutrino participates only in the weak interaction, while a quark participates in all three. The gluon and the photon are massless and travel at the speed of light, while the W and Z bosons are always bound to a Higgs particle, consequently becoming massive and slow. The photon and the Z boson are uncharged, in contrast to the W boson and gluon, both of which are charged. The gauge force mediating between two electromagnetically charged particles diminishes with separation distance, while the strength of the strong force increases with particle separation!

Now we examine the properties of quarks to illustrate the nature of the relationship between the charge of a particle and its bearing on the gauge interaction. The color charge of a quark comes in three varieties, referenced as red, green, and blue. Color charges are analogous to the more familiar two varieties of electromagnetic charge, positive and negative. The color charge is confining, as it causes three quarks of different color, along with the color-charged gluons they exchange, to aggregate into a composite structure called a nucleon (proton or neutron). The field of gluons extends for a short distance beyond the structure of the nucleon, giving rise to the short-ranged residual strong force, which has the effect of binding protons and neutrons into atomic nuclei. Quarks also possess properties imposed by the global symmetry sector (their flavors, which take either an up or a down value) that contribute to the characteristics of the nucleon they comprise. Thus, a nucleon having 2-up plus 1-down quarks constitutes a proton, while a neutron consists of 2-down plus 1-up quarks. Since each of the three quarks also possesses an electromagnetic charge, their nucleonic superstructures must display those properties collectively as well; for instance, up quarks have an electrical charge of +2/3, while down quarks have a charge of -1/3. Therefore, a proton must have a net electrical charge of +1, and a neutron, a net electrical charge of 0, as observed.

The confined internal symmetry space of a particle is small, but it is not infinitesimal, so that in microscopic regimes, point-based field theory characterizations must give way to the inherent details of its very active internal space. To this point, we have considered only the gravitational metrics and connections in the external symmetry space. Metrics and connections with dynamic topological transformations are also native to the confined internal symmetry spaces of particles. In Chapters 13 & 15, we probe more deeply into the internal symmetry spaces of elementary particles using the devices of string theory, which introduces finite-sized vibrating strings, undulating braneworld surfaces, and six-dimensional, metric-bearing Calabi-Yau manifolds.[247]

Dynamism within the internal symmetry space of a particle demonstrates the important discovery that the energy and power derived from gauge symmetries reaches into the very fabric of the macrocosm. In our case-study of the strong interaction, three quarks naturally integrate to compose the confined symmetry space of a nucleon, which, in turn, exhibits its own informational properties with respect to the coevolving macrocosm. As a result of this prototypical occurrence, a scale-free fractal evolution of structure commences. Nucleons bond into atomic nuclei as the result of the residual strong gauge force, while still signaling the unconfined, net positive electromagnetic charge of their composite quarks. The nuclei, in turn, bind electromagnetically to the negatively charged electrons ambient in the charge-symmetric universe, thereby constructing the chemical elements, which are the ingredients for the self-organization and evolution of complex systems. This generative fractal order extends from the tiniest particles, to the composition of the human brain, to societal organization.[248]

Although the internal symmetry spaces of nucleons, atomic nuclei, chemical elements, and compound molecules sow the seeds of integrated information into the anthropic pathway, they are inadequate to represent non-trivial objects of consciousness. These basic compounds are too rudimentary to support the hard-coded process cycles necessary for the *adaptable* generation and storage of inter-causal information. They are essentially, in their stable forms, isolated systems with limited (semiotic) interaction versatility. In the metal-rich environment of early earth, where an extensive variety of abiotic electrochemical atoms and molecules was available, many bonding alternatives that consumed the same reactant resources coexisted. The formation and dissolution rates for the various chemical products settled into a natural balance with respect to their bonding affinities and relative concentrations. In conditions of limited resources, complex Lagrangians evolved which reflected the strongest of the interaction *networks* governing balanced ingredient processing. This mechanism established webs of chemical pathway interdependencies and the emergence of stable, but flexible, sub-network structures; that is, adaptable, open-system microcosms, coevolving with the environmental substrate in a relationship based upon gauge symmetry interactions. In favorable environmental niches, these individuated microcosmic systems were able to increase in complexity and adjust to external pressures, heralding the onset of biological natural selection.

Thus far, our discussion of gauge symmetries and their relationship to emergent, information-bearing structures and processes has been limited to classical physics. In the classical domain, gauge force-fields mediate interactions among elementary particles and their composites. But these represent only one aspect of the gauge symmetries, albeit the aspect most immediately pertinent to the neural process-structures of vehicles of consciousness. There is a higher gauge symmetry in the macrocosm that encompasses these classical effects, as well as the realities of quantum physics that underlie the origins and foundations of the classical regime. To illustrate, let us consider the full gauge symmetry of electromagnetism encoded in the electro-magnetic *potential (Φ-A)*, as opposed to only the classical electro-magnetic *force fields (E-B)*. The classical force fields of electromagnetism (which are transmitted through space via photons; i.e., the field disturbances which vary in strength as a function of the spatial location and velocity constraints of relativity) may be derived from the electromagnetic potentials, but not vice versa, as the potentials have a higher, redundant gauge freedom than the fields.

The gauge symmetry of the electromagnetic potential implies unfamiliar, non-classical phenomena such as *nonlocal* correlations, or entanglements, among defined subsystem entities or events. Let us clarify our use of the terminology since it is conflated in the literature of theoretical physics. The term nonlocal refers to kinematical features of a system apart from the dynamical interactions among sub-systems *defined* locally, which are constrained by special relativity to have a subluminal communication messenger, such as a field or particle.[249] The term nonlocal is often used to describe a system that would violate Einstein locality, and therefore relativity, but we take nonlocality to describe the *non-separable* physical relationships and interactions that have been shown to exist independently of predefined spatial locations and velocities.

> *When measurement events on entangled systems are spacelike separated, parameter independence (unlike outcome independence) is a consequence of relativity theory... all fundamental particles can be considered to be excitations of underlying non-pointlike entities in a multi-dimensional space. The particles' intrinsic charge, mass and spin may then arise as nonseparable features of the world at the deepest level.*
>
> — Stanford Encyclopedia of Philosophy[250]

The interpretation that entanglement is a global effect is also imprecise. Global means effective at all points in a Lorentz-broken spacetime, which is not to be imposed on entanglement correlations. It has recently been conjectured that quantum entanglement and gravitational wormholes are unified by the Lagrangian of a deep-seated symmetry principle below the threshold of spacetime itself, which invites the nonlocal ontology.[251]

Nonlocality, despite Einstein's characterization of it as "spooky action at a distance," has been verified by numerous experiments, beginning with the demonstration of the Aharonov–Bohm effect in 1960,[252] and recently by the first "loophole-free" experiment in 2014.[253] "This effect is now recognized to be of great theoretic importance as it is the first example of quantum gauge phenomena." [254] In the Aharonov–Bohm experiment, a changing magnetic flux that was totally isolated within the confines of a solenoid coil was able to alter the pattern of interference being produced on a detector screen by two electron beams outside of the solenoid. The means by which the magnetic flux altered the interference pattern on the screen, and therefore the relative phases of the interfering electron beams, has to do with non-classical *active information,* as termed by David Bohm in his ontological interpretation of quantum theory. Active information is a kind of subtle, guiding information, inhering at the deepest levels of physical reality. In the primordial universe, preceding the cascade of classical sub-wholes and gauge force fields, the parent quantal symmetry ensemble encompassed the full potentiality of active information.

* * *

14

Enfolded Symmetries

Symmetries can be distinguished as *continuous* or *discrete.* Rotation is a continuous symmetry because, according to the symmetry Lagrangian of classical physics, there is no limit to how infinitesimally small an angle through which a physical object may be rotated. As a concrete example, the rotation of the planet earth about its axis of spin is a continuous symmetry. In contrast, the direction of the spin axis of a quantum particle, relative to the direction of its motion, is a discrete symmetry, called parity symmetry. There are only a finite number (two) of representations in the ensemble of the quantum particle's spin axis symmetry space, left- or right-handed. Symmetries can also be *exact* or *broken.* An exact symmetry is unconditionally valid throughout an entire symmetry space, whereas a broken symmetry is valid only for a subgroup of that symmetry. The so-called "spacetime symmetries" of rotation and parity, as well as Lorentz invariance (special relativity) and general covariance (general relativity), are examples of exact symmetries.

Broken symmetries are connected with the very essence of physical manifestation. Symmetry breaking actualizes the representations of the many symmetry spaces we encounter in our daily lives. Even events such as selecting a card from a deck invokes symmetry breaking. Before the selection of a card, the complete symmetry space of possibilities includes 52 representations. Once a card has been selected, however, the symmetry space is broken down to the subset of one possibility, the manifest card. An example of symmetry breaking in modern physics is the (weak-force) quantum spin of a neutrino, which is always measured in experiments to be representing the left-handed spin subset, although right-handed spins are mathematically consistent with parity symmetry (in contrast, the spin parity of quarks is unbroken). In this example, the symmetry of the "law" for neutrino spin is *explicitly* broken, hence the Lagrangian of the standard model has been adjusted by the substitution of an asymmetrical broken symmetry term for the neutrino, as opposed to the exact symmetry term describing the behavior of a quark.

While studying the electromagnetic and thermal properties of crystal structures, Pierre Curie realized that the phenomena observed were a function of the *asymmetry* delineating the lattice structure (relative to isotropic three-dimensional space). The manifest phenomena occur in a reduced symmetry subspace, which breaks the larger symmetry: "The phenomena break the symmetries of laws."

Sir Isaac Newton was sitting under an apple tree one fortuitous afternoon when an apple fell and hit him on the head. This legendary encounter inspired Newton to conceive his theory of gravity. Two different kinds of broken symmetry are involved in this anecdote. The gravitational gauge force symmetry is explicitly broken by the mass of the earth into a subset in which all trajectories point in the downward direction, hence guiding the famed apple from the tree to its celebrated destination. This is the broken symmetry of concern to Newton's theory. But gravity, of itself, is a representation of a different broken symmetry, one that was broken spontaneously from the supersymmetry of the big bang.

Spontaneous symmetry breaking affects global symmetry spaces across domains of nature, ranging from a bar of magnetized iron, where all constituent atomic spin axes are parallel, to the entire universe, where the spin axis of a neutrino is exclusively left-handed. Global symmetries are generally broken by the change of an ambient condition, often decreasing temperature. The term "spontaneous" means that the symmetry is not broken by particular interactions but rather by the global asymmetry that appears uniquely in the lowest energy (or generally, extremal) state of a given parent symmetry ensemble. This state is referred to as the *vacuum*, or *condensate*. Cosmological events of spontaneous symmetry breaking occurred sequentially in the very early universe as temperatures cooled below condensation thresholds, enabling the types of matter to crystalize. Each successive stage of cosmological symmetry breaking resulted in two partitions of the parent symmetry: a broken condensate, and a residual, unbroken subset of the parent symmetry. Each of these partitions appears as an effective action term in the resultant Lagrangian; in the framed standard model, separate terms for the broken and the residual unbroken symmetry partitions are included for each of the four symmetry groups.

Spontaneous symmetry breaking describes the condition in which the laws of a theory are invariant, but manifest systems are not, because the vacuum condensate to which the systems couple is non-invariant; i.e., the vacuum state is unique in that it does not respect the symmetry of the theory. The vacuum condensate of a spontaneously broken symmetry both carries potential energy and confers global order, called the order parameter. The familiar example of ferromagnetism demonstrates the concept of spontaneous symmetry breaking. When the temperature of an iron bar drops to the Curie temperature, the randomly oriented spin axes of its constituent atoms spontaneously rotate into parallel alignment, giving the iron bar a collective polarity, called its magnetization, or order parameter, which becomes the background for emergent collective spin wave excitations. The residual symmetry laws of the unbroken sector govern the dynamics of the spin-waves (phonons), while the broken condensate sector acts to attenuate the velocity of the spin-waves. Reheating the iron bar sends the atomic axes back into the randomized state of the full unbroken symmetry, where no particular Lagrangian solution for a given atom's orientation is imposed.

In quantum field theory, both the broken symmetry and residual exact symmetry sectors are viewed as continuous spacetime-filling fields in which local manifestations arise in the form of standing wave packets, bearing the quantal properties that define the species of particles. The residual unbroken symmetry fields are the matter and gauge fields, while the broken symmetry fields are ordered background condensates. All of the varied broken and unbroken fields comingle in such a way that stable excitations appear which exhibit the characteristics of particles, including their spin, charge, and mass. In general, numerous cross-correlated fields interact and jointly constitute the substratum for the manifestation of nature's plethora of particle species and interactions. [xi]

* * *

Based on the observed rate of expansion of the universe, from our time-asymmetric perspective, a big bang occurred at some finite time in the past. According to the standard model of cosmology, the universe has been expanding for 13.8 billion years since that primordial time, when it was a hot, dense plasma. Many scenarios have been crafted to describe the

---

xi Since the big bang, hierarchical diminution of symmetry group dimensionality has descended through the groups symbolized by E8xE8 ~ SO(32) ~ SO(10) ~ [R4xSO(3,1)]xSU(5) ~ [R4xSO(3,1)]xSU(3)xS(2)xU(1), with each successive group having differentiated, or "broken," out of the prior. This means that the physics of the primordial E8xE8 universe of integrated supergravity-supersymmetry has now broken down to, and released, electromagnetism, whose U(1) internal symmetry space involves only a phase factor. SO(10) accounts for the 4 dimensions of spacetime plus the 6 compact internal dimensions parametrizing the standard model gauge symmetry. The *particle species* are the irreducible representations of the 4-dimensional Poincare spacetime symmetry group (symbolically denoted by the group R4xSO(3,1), where R is translation in 4-dimensional spacetime, and SO(3) is Lorentzian rotation which holds the velocity of light as a constant). The Poincare group includes vector bosons (gauge field carriers), spinor fermions (individuated particles), and a scalar boson (the Higgs condensate). The Lagrangian during the supergravity-supersymmetry epoch is formalized by the super-Poincare algebra which contains high-level symmetry transformations that turn bosons into fermions, and vice versa.

big bang and its dramatic consequences. Some theories posit a singular event which produced the cosmos from a point of zero size and infinite energy density. These models invariably lead to inconsistencies between the two great advances of 20th century physics: general relativity applying to the macroscopic universe, and quantum field theory pertaining to the microscopic. Since the 1980's, conceptually broader theories have been developed, including string/M-Theory, loop quantum gravity, imaginary time, and conformal cosmology, each of which circumvents these inconsistencies by allowing possibilities such as holographic transitions and cyclic aeons of manifestation. In the mainstream of the contemporary cosmogonies, whether singular, imaginary, or cyclic, all agree upon three principal retrodictions; since the big bang event there have ensued: the expansion of space, the progressive breaking of symmetry, and the condensation of matter into structure.

At the big bang, the universe was in a state of (asymptotically) complete, but unstable symmetry, which caused a series of symmetry breaking transitions to ensue. One of the first condensates to settle-out was the vacuum condensate of *negative* "dark energy," postulated to explain the observed *expansion* of space. In the nascent supersymmetric universe, all particles (transient excitations) shared many degrees of freedom and readily interchanged their identities. In this state, physical entities, as fundamental as time, charge, and spatial parity, were symmetrical and thus interchangeable. Then, within the first instants of the cosmic cycle, cooling, induced by dark energy expansion, caused a series of further symmetry breaking transitions and consequent restructurations of the universe. The major transitions through which the macrocosmic universe has evolved are shown in Table 14-1. [xii]

---

xii *Engineering Notation* is used in Table 14-1 for readability of the powers of 10. Hence, 10 to the fourth power = 10E4; and 7.2 x 10 to the minus forty-third power = 7.2x10E-43.

Time (sec.)	Temp (^{0}K)	Event	Notes
0		Big Bang	Expansion of universe begins
10E-43	10E32	Plank epoch	Super quantum potential only
10E-40		Theory of Everything (ToE)	Supergravity: R4 x SO(1; 3).
		Gravity and standard model break	First excitation foliation
		Hidden and visible branes	Dark matter
10E-36	10E28	Grand Unification Theory (GUT)	Supersymmetry: SU(3) x SU(2) x U(1)
		Copious particle production	Photons, quarks, and leptons only
		Strong and electroweak break	Higgs field turns on
10E-25		Radiation domination	Relative to matter and dark energy
		Electroweak epoch	CPT asymmetry
10E-11	10E16	Electroweak phase transition	Weak and electromagnetic break
10E-6	10E13	Hadron epoch	Protons and neutrons form
1		Hadron threshold	Hadrons annihilate
3		Lepton threshold	Leptons annihilate
10		NBR	Neutrinos decouple
10E3		Big Bang nucleosynthesis	Light nuclei form
(years)			
10E5	3x10E3	Matter domination	
3.8x10E5	2.73	CMBR	Photons decouple, atoms form
0.4x10E9		Galaxies	Galaxies and quasars form
7.5x10E9		Our solar system	
8.2x10E9	300	Anthropic sector	Life on earth
9.8E9		Dark energy domination	

Table 14-1 *Cosmic Cycle of the Macrocosmic Symmetry Space*

While the temperature of the universe was still above 10^{28} degrees Kelvin, within 10^{-36} second after the big bang, a highly symmetrical Lagrangian was in effect, unifying diverse features of the universe in the so-called "Theory of Everything." In this epoch, supergravity and the standard model supersymmetry were unified; that is, the gravity particles (the quantized field excitations of curved space) and the particle excitations of matter and their gauge fields were interactive and interchangeable. Boson and fermion particle identities, charge conjugations (matter and antimatter), and time itself all transmuted freely.

This unintuitive "identity crisis" of particles actually brings some sense to the minuscule numerical values that are calculated, according to big bang theory, for the time intervals that separated the major transitions of the early universe. For example, the symmetry space described by the Theory of Everything, which is calculated to have existed in the epoch from the "time" of the big bang until only 10^{-36} second later, dissolves time into the state of unified supergravity-supersymmetry. The understanding that time itself was convoluting with other dimensionalities in the rapidly inflating embryonic universe helps to reconcile the astonishingly tiny numbers calculated for the time intervals that elapsed between major physical epochs in the aftermath of the big bang.[255]

Continued cooling of the universe induced the unified supergravity-supersymmetry of the Theory of Everything to break spontaneously into two partitions. In the supersymmetry partition, described by the Grand Unified Theory, the strong nuclear, weak nuclear, and electromagnetic interactions were initially unified, but in another round of symmetry breaking, the strong nuclear and electroweak interaction gauge fields, and particle charges, settled out. However, the electromagnetic gauge force, carried by the photon gauge field, had not yet separated from the parent electroweak symmetry; all gauge interactions in the electroweak sector during this epoch were mediated by X-bosons. Then, 10^{-11} seconds after the big bang, electroweak symmetry breaking occurred, producing four gauge bosons in the residual sector, called the W+, W-, Z bosons, and the photon.

The W+, W-, Z bosons attained mass through interaction with the Higgs condensate, which had condensed concurrently in the broken symmetry sector. In contrast, the sibling photon did not share a dimension with the Higgs condensate, and therefore remained massless. Referring to our previous discussion of the framed standard model, we recall that it is the global symmetry sectors which are broken, in this case producing the Higgs condensate, which reacts with the W+, W-, Z bosons via the weak force; while the electromagnetic gauge photon retained its freedom in electroweak symmetry breaking.

The Lagrangian of supergravity-supersymmetry theory includes in its symmetry space not only the gravitational gauge boson, the graviton, and the palette of elementary bosons and fermions listed in the "minimally symmetric standard model," such as the photon, quark, electron, etc., but also a *superpartner* for each of these particle species (none of which have yet been observed). Superpartners are identical to their standard model counterparts except for the value of their spin, which determines their *behavior toward other particles*. Particle behavior falls into two categories: integer spin-valued bosons, which collectively integrate, and half-integer spin-valued fermions, which maintain autonomy.

The observable properties of bosons and fermions manifest in integer multiples of discrete quantum numbers. This quantized behavior reflects that it is the conjugate square of a particle's *inferred* internal properties, such as spin composition, that registers on measuring instruments. The spin composition of a particle is not directly observable, but the resultant gauge effects imply an inner symmetry space with complex-valued mathematical properties.

The cooling of the universe and the progressive breaking of supergravity-supersymmetry led not only to the formation of vacuum condensates and emergent residual field interactions, but also to other important changes in the organization of the universe. By the time the ambient temperature had dropped below 10^{16} degrees Kelvin, one member from every super-partner duet was too massive to participate in further interactions. These more massive partners, some bosons and some fermions, were unstable at the lowered temperatures and immediately decayed into neutralinos, the lightest possible supersymmetric particle.

Neutralinos interact very feebly with respect to the standard model forces and our devices are as yet unable to detect them. Furthermore, neutralinos do not participate in the electromagnetic interaction and are therefore invisible to photon detectors, including optical telescopes on earth and the Hubble satellite). However, the predicted quantity of remnant neutralinos is enormous, and though neutralinos are extremely light, their aggregate gravitational effect is significant. Neutralinos have been inferred, in some theories, to explain the origin of the dark matter which would account for observed anomalous inter-galactic dynamics, as discussed in Chapter 15. (Small black holes, having primordial origins, possibly associated with the massive superpartner sector of the full standard model symmetry, is also a currently popular proposal for the composition of dark matter.)

Another effect of the expansion of the universe was that the quantum fields became too spread out and weak to produce the high amplitude mass-energy fluctuations necessary for the production of stable, standing wave packets, or "real particles." Below the mass-energy thresholds required for particle production, however, the underlying quantum fields continue to exist. These fields contribute a zero-point energy to the vacuum, even in the absence of material particles (relative to a stationary inertial frame). The zero-point energy of the vacuum exhibits quantum properties that imply a sea of "virtual particles."

Virtual particles cannot be directly observed, but their effects can. For example, the Casimir effect causes parallel plates to draw together when they are separated by less than the "size" of a virtual (or real) particle. This is due to the pressure of the large-wavelength wave-packets (virtual particles) fluctuating outside of the plates, but not inside, because they cannot fit. Further, the charge and spin polarities of sub-threshold virtual particles have manifest consequences. The virtual electromagnetic field has been empirically demonstrated to influence the motions of electrically charged real particles, such as electrons and protons; and the Unruh effect causes real particles to emerge from the vacuum when an observer accelerates through it, compliant with the equivalence principle of general relativity. Virtual particles are an essential component of the Lagrangian of quantum field theory.

At the macrocosmic scale, spontaneous symmetry breaking characterized the evolution of the universe. At the microcosmic scale, it was crucial to the evolution of life within the symmetry space of earth 4.5 billion years ago. When the ambient temperature of earth cooled sufficiently, atmospheric steam underwent spontaneous symmetry breaking by condensing into the liquid phase of water, providing a requisite ingredient for the formation and survival of organic life.

Another critical effect of the spontaneous symmetry breaking of water occurs continually on earth, when it undergoes a phase transition from the liquid state to solid ice. In cold temperatures, water molecules are able to align to their least-energy state, due to thermally less-perturbed weak hydrogen bonds becoming stable enough to sustain a crystalline structure. In the abnormal case of water, the solid state has a lower density than the liquid. This enables a crust of ice to float upon basins of liquid water, thus insulating and protecting aquatic life from lethally frigid conditions.

Symmetry breaking alone does not account for the manifestation of complex systems. On earth, the kind of spontaneous symmetry breaking that characterizes water and ferromagnetism is accompanied by another type of symmetry change, which is driven by the process of natural selection.[xiii] As described in Chapter 3, the selection-invention process is pervasive throughout the universe, ranging from the quantum regime to the sociologic. In particular, natural selection on earth has fashioned the symmetry spaces of living systems. These emergent holistic symmetry spaces have integrated and evolved to the very high dimensionality of the human brain. But necessarily, the exact microcosmic symmetry spaces available to natural selection become broken as well. For example, in all living organisms, certain types of helical molecular structures are left-handed only, and although right-handed configurations could exist according to the laws of chemistry, and have been synthesized in the laboratory, only left-handed molecules have been observed in nature; analogous to the exclusively left-handed neutrino spin observed in the macrocosmic symmetry space.

The biosynthesis of the left-handed versions of organic molecules was selected early in evolution for the advantage of process consistency within and among interdependent holistic organisms. The symmetries broken by natural selection are abundant in organic life systems, as well as in their reciprocative biospheres, and display geometrical symmetries such as molecular parity and chiral (handed) autocatalytic cycles. Thus, while life systems have risen on emergent platforms of exact symmetries, reflecting nature's fundamental foundation, the branches of evolution have manifested through optimized selection of broken symmetries.

---

xiii *Natural selection* is the process by which the *natural* environment of an organism *selects* for (interacts with) those traits of the organism which confer upon it a replicative advantage, so that it eventually comes to dominate a population.

The numerous symmetries enfolded within the supersymmetry of the universe include countervailing exact and broken sub-symmetry sectors, in both the macrocosmic and microcosmic domains. The relative influence of each of these sectors is dependent on prevailing conditions, such as position in the cosmic cycle, ambient temperature, and available resources. The exact symmetry sector may be further classified into three types: the residual, restored, and emergent symmetries.

The *residual symmetries* are those subsets of a parent symmetry that have survived a symmetry breaking transition, such as the SU(3) symmetry of the strong nuclear interaction. The *restored symmetries* result from a local return to certain ambient conditions, usually high temperature, for example the restoration of the SU(2) x $U(1)_y$ electroweak symmetry taking place in the hot core of a condensed star or in a high-energy collider. The *emergent symmetries* are described by Lagrangians that govern the generative orders of causal correlations, including the vertical/horizontally crisscrossed interaction networks of fractally unified classical structures, the self-organizing complexes of holistic microcosmic organisms, and the integrated, centrosemiotic, phenomenal symmetry space of the human interpretant.

Of special importance to biological systems is the U(1) symmetry group, which mathematically codifies the Lagrangian of electromagnetism, the interaction through which electrically charged chemicals evolve into complex, living, phenomenological, process-structures. Interestingly, the particular $U(1)_{em}$ group which specifies the electromagnetic interaction, and all activities of its descendent biological processes, happens *not* to be a fundamental field of nature in the sense of initial cosmic origin, but rather arose indirectly as a byproduct of electroweak symmetry breaking. Mathematically, however, electromagnetism *is* fundamental in that it expresses the general property of all U(1) symmetry groups; namely, complex-number valued transformations. In the instance of electroweak symmetry breaking to the U(1)em subgroup, the complex-valued wave-like behavior of photons became independently effective.

Once the universe had cooled sufficiently and thermal agitation had subsided, the short-range forces of the strong and weak interactions were able to confine stable composite structures, locking quarks into nucleons, and nucleons and electrons into the atomic elements. Effectively, the long-range gravitational and electromagnetic potentials were liberated to function independently of the nuclear forces now sequestered within tiny, self-neutralizing, composite nuclear structures. The relatively weak gravitational and electromagnetic forces were thereby enabled to coalesce and bind these elements into stable aggregations, such as planets and molecules, respectively.

On earth, the gravitational field has assumed the role of an explicitly broken environmental condensate polarized into up and down vertical vectors. Nevertheless, gravity is instrumental in many non-vertical processes consequential to life, for example: in the environment—powering the water cycle of the troposphere; in the interface between organisms and environment—aquatic vertebrae selecting for bilateral symmetry to expedite head-first locomotion parallel to the surface of the earth as they forage in their gravitationally flattened environments; and in the human nervous system—selecting for a keen vestibular sense, as exemplified by gymnasts performing their magnificent routines on the balance beam.

In the tranquil molecular environment of deep-sea vent chambers, electromagnetic fields—unfettered by the effects of confined nuclear forces—were free to mediate interactions among the chemical building blocks that had come together. This environment provided the platform upon which feedback and feedforward reaction cycles self-organized into sets of interdependent macromolecules. Organic polymers such as RNA and peptides grew into complex, cross-connected, twisted and folded assemblies, held together by faint but unperturbed hydrophilic/hydrophobic, hydrogen, and the Van der Walls forces, establishing a domain for newly enfolded symmetry transformations. The evolution of complexity within this symmetry space required an external environment that provided the chemicals necessary to reinforce the autocatalytic cycles underlying the self-organization of unified, open-system, dissipative structures. On metal-rich earth, this resource requirement was fulfilled, and the closure of autocatalytic sets was attained, endowing emergent holistic microcosms with the ability to sustain and perpetuate life.

* * *

15

The Macrocosm

The *braneworld* model of physics is based on the idea that the particles of nature are fundamentally vibrational in character and that their spectra of vibrational modes, or tones, endow them with properties sufficient to satisfy symmetry Lagrangians comparable to those of the empirically successful field theories of general relativity and the standard model. The simplest case of a vibrational element is a one-dimensional physical object called a *string,* whose modes of vibration are analogous to the overtone notes of a guitar string. Similarly, the undulating membrane of a drum portrays a two-dimensional *brane,* or 2-brane, and the three-dimensional space in which we live is a 3-brane.

Branes with spatial dimensionalities greater than three also play a role in physics. According to M-Theory,[256] once the process of expansion and cooling had been initiated immediately following the big bang, an eleven-dimensional symmetry space underwent a series of spontaneous symmetry breaking events, producing a cascade of lower-dimensionality brane condensations that were superimposed by residual quantum gauge fields and vibrational objects.

Multi-dimensional branes may overlay and project through one another. In the supersymmetry-broken sector of M-Theory, the four-*extended-dimensional* macroscopic spacetime of general relativity (3-brane plus time) is overlaid by the six-*compact-dimensional* microscopic manifold of the internal symmetry space of standard model particles; i.e., at every point in four-dimensional spacetime there potentially exists a curled-up, six-dimensional, finite-sized manifold enwrapped by undulating branes whose characteristics account for the properties of particles. Combined, these ten dimensions constitute the unique platform that is compatible with the five mathematically possible superstring theories, supplemented by a sixth, low-energy supergravity theory.

M-Theory morphs the six lower dimensional theories into one another as a function of abstract moduli, which are so fundamental as to underlie the emergence of spacetime itself. The six sub-theories unified by M-Theory are related through a web of dualities that associate the limiting conditions of the symmetry space of the M-Landscape. For example, the T-dualities smoothly transition small and large distance scales, as well as the radii of dimensional cycles and the winding numbers of their enwrapping branes. The S-dualities are effective for strong and weak coupling strengths, as well as these coupling strengths and the dimensionalities.

The braneworld paradigm is welcome to physics. Conventional field theories are problematic in high energy regimes, and for nearly a century they stifled progress toward the unification of relativity and quantum theory. Field theories are generally set upon a continuous manifold of infinitesimal points, which constitutes the background space where the elements of a theory live and interact. Actual physical manifestation in nature, however, does not behave continuously in the smallest spacetime regions, but is quantal instead, happening in specific non-zero sized actions that are mathematically incompatible with dynamical continuity.

Based on our everyday experiences in the coarse-grained environment of earth, it is natural for us to conceive of physical reality as unfolding in a spatiotemporal continuum of points, but generalization of the idea of a continuous spacetime that applies at all scales is not only technically unjustified, but imposes an epistemological constraint on theories such that some of nature's features cannot be adequately expressed. Even in classical general relativity where the laws are set upon a spatiotemporal continuum, Einstein demonstrated with his "hole argument" [257] that it is meaningless to consider a region of spacetime in the absence of contained matter. Consequently, another ontology is needed to supersede the only contextually effective continuous physical field theories of relativity and the standard model. A vibrational model is required to remove the mathematical singularity breakdowns of field theories so that the *sounds* of reality may be heard.

Of primary concern to big bang cosmogony is the implication that the universe must have initially held the potentiality for the generation of order, as it exists throughout the universe today. This seems to conflict with the observation that the universe as a whole has been evolving toward a state of disorder since the big bang, the statement of which is the second law of thermodynamics. Astronomical evidence indicates that the universe is expanding and settling into a spread-out state of disordered, high-entropy thermodynamic equilibrium, akin to a gas that has escaped from a container and expanded into a diffuse, unorganized state (inside a black hole, entropy increase signifies a rise in gravitational saturation). The theory of *conformal cyclic cosmology*[258] accounts for the reservoir of potential order that must have existed in the big bang era (through invocation of the principle of *conformal symmetry*—invariance of conformation, or shape; and independence of size). In a conformal big bang transformation, the residual uniformity characterizing the end of an expanded, equilibrated cosmic cycle, is holographically inverted at the beginning of the next cycle into the potentiality for order that is latent in the homogeneity of the budding universe.

During a cosmic cycle, physical space expands and flattens, causing its curvature and associated gravitational degrees of freedom to decay. This effectively reduces the volume of the late universe's phase space (the symmetry domain of all possible states an isolated system can occupy), with the result that its phase space is conformally transformed through the so-called "big crunch." In terms of thermodynamics, a reduction in entropy is tantamount to an increase in order. As the cosmic cycle proceeds through the big bang transformation into the small-scale state of the forthcoming cycle, the potential for the formation of orderly structures and systems becomes available due to the relationship between gravity and the ambient homogeneity (translational uniformity) and isotropy (rotational uniformity) of the early universe.

The existence of homogeneity and isotropy in the initial epoch of our current cosmic cycle is supported by the extreme uniformity observed in cosmic microwave background radiation. This background radiation is comprised of the photons which, due to lowered temperature 300,000 years after the big bang, decoupled from the proton-electron interaction which had suddenly stabilized through the bonding of electromagnetically neutral atoms. Since decoupling, these photons have traveled freely throughout the expanding universe. Recent mapping of the relic background radiation discloses the uniformity of the early universe, and thus the gravitational potentiality for the emergence of thermodynamic order, complexity, and information.

The terms *order, information, data, entropy, and complexity* refer to concepts crucial to understanding the nature and evolution of consciousness, but explanations for what these terms actually mean differ greatly. As illustrated in the footnote, ambiguous and contradictory meanings for these words are commonplace in the English language.[259,260] Even within specialized disciplines of science, use of these terms is generally contextual and may or may not presume a subjective aspect.[xiv] However, in light of

---

xiv *Common Definitions Related to the Concept of Order.*

Order

- The arrangement of objects in position or of events in time.
- A system functioning according to some definite laws or rules.
- The totality of arrangements composing some sphere of action or being.

Information

- Something *received or obtained* through informing: knowledge *communicated* by others or obtained from investigation, study, or instruction.
- A numerical quantity which measures the *uncertainty* in the outcome of an experiment to be performed. When an event occurred whose probability *was* p, that event is said to have *communicated* an amount of information ~ (log $1/p$).
  - *Knowledge* which is gathered as a result of processing data.
  - *That which informs*, i.e. that from which data and knowledge can be derived.

Data

- Something that is given from being *experientially* encountered, or from being admitted or assumed for a specific purpose: a fact or principle that is granted or presented.
- Something upon which an inference or argument is based, or from which an intellectual system of any sort is constructed.

Entropy

- (Thermodynamics) On the macroscopic scale: a function of thermodynamic variables, such as temperature, pressure, or constituency, which measures the energy not available for work during a process.
- (Statistical mechanics) A measure of randomness and disorder among the microscopic constituents of a thermodynamic system.
- (Data transmission and information theory) A measure of the loss of information in a transmitted signal or message. (Physical information without meaning.)
- (Cosmology) The tendency for the universe to attain a state of maximum equilibrium and homogeneity, in which all matter is at a uniform (macroscopic) temperature.

Complex

- Having many *interrelated* parts, patterns, or elements—complicated is to simple as complex is to independent.

the consistent model that has recently consolidated in the sciences, we may now ascribe operational definitions which standardize the meanings of order, information, entropy, and complexity across the disciplines. *Order* applies to a set of correlated elements that demarcate a whole, such as the correlated pattern of ice crystals that configures a snowflake. In contrast, when elements exhibit no correlations, such as clothes strewn about a room, or molecules of a gas flying and ricocheting randomly within a closed container, there is a state of disorder.

The second law of thermodynamics is an observed property, or tendency, of nature. It states that for any isolated system, an *abstract* quantity called entropy invariably increases when a change occurs among the parts of that system. The ensemble of all possible configurations that a system can occupy (phase space) may be partitioned into groups of states having a particular measurable macroscopic value for a certain attribute of the system as a whole, such as temperature, pressure, density, color, or chemical composition. The number of states having a particular value, defining the "volume," V, of that group of equivalent states, determines the state's entropy, S, according to Boltzmann's equation: $S=k \log V$. The Boltzmann volume, and thus the entropy, for the state of a system are a function of the number of constituent parts in the system, as well as the number of ways the system can yield a particular macroscopic value.

In the oft-cited example of a gas in thermal equilibrium inside a closed container, the (log of the) number of possible molecular arrangements corresponding to a given temperature is equal to the entropy of that system. Since there are many more ways that a system of water molecules can be configured when boiled into steam diffusing randomly through space than there are for those molecules to be configured into the frozen pattern of a snowflake, the entropy of the disorderly gaseous steam is therefore greater than that of the orderly snowflake. The correspondence between the second law of thermodynamics, the entropy of a system, and the properties of order and disorder was thus believed—since the mid-nineteenth century assertion by Rudolph Clausius[261] with respect to heat engine thermodynamics—to be a sacrosanct law of nature. In certain circumstances—gravitational degrees of freedom (Chapter 2), dissipative structures (Chapter 3)—it is now understood that this intuitive connection between entropy and disorder breaks down.[262] The terms order and disorder are gradually being pruned from scientific vernacular describing entropy. The *dispersal or diffusion of energy to, or the occupancy of, more microstates* characterizes an increase of entropy.

The early universe consisted of photons and matter coexisting in thermal equilibrium—a state of homogeneity and isotropy signifying maximum thermodynamic entropy. As a result of this uniformity, gravitational forces which were symmetrical in all directions cancelled mutually, so that the potential gravitational degrees of freedom in the supersymmetry Lagrangian were inactive. In the cooler temperatures that ensued, stable matter began to decohere and clump together, initiating an emerging gravitational field wherein the clumps of matter tended to attract, rather than disperse. As the gravitational degrees of freedom became activated, the formation of stars and galaxies commenced. The "second law of entropy increase," as applied to gravitation, leads not to a state of diffuse equilibrium, as in thermodynamics, but to a state of maximum entropy distinguished by coalesced matter saturation, which is attained within the horizon boundary of super-massive black holes residing at galactic centers.

The activated gravitational degrees of freedom resulting from localized clumping of matter increased the volume dimensionality of the universe's phase space relative to the uniform thermodynamic baseline set by big bang homogeneity/isotropy, and consequently established the potentiality for entropy increase. Correlated states were able to assemble themselves into far-from-thermodynamic-equilibrium or gravitational saturation, and evolve into ordered, information-dimensional symmetry spaces. A unifying definition of entropy applying to both thermodynamic systems and black holes is: "A measure of hidden information, often the information stored in things too small and numerous to keep track of" [263] (where "information" refers strictly to physical quantity, i.e., Shannon information, and not to quality or meaning relative to an interpretant; entropy does not entail semantic connotations).

The gravitational attraction of matter that caused stars to form persisted until thermal collisions intensified within their cores, triggering nuclear fusion reactions that emit high energy photons into space. In particular, photons from our sun are absorbed by matter on earth and reemitted into space at lowered temperatures. Since energy is conserved in this process, there must be a greater number of cool-emitted photons than hot-incident photons. These additional photons represent degrees of freedom that increase the phase space volume of possible state configurations on the surface of the earth, thereby opening a window of opportunity for the emergence of ordered systems. These emerging degrees of freedom are captured for the building of the chemical structures and processes, such as photo-synthesis, which are essential for the construction of life systems.

Cyclic braneworld theories describe a universe that springs into action with the condensation of two ten-dimensional branes, connected by the unifying eleventh dimension of supersymmetric M-Theory.[264] Figure 15-1 depicts the braneworld cycle of the macrocosm (cycles of evolution are also postulated in contemporary quantum field cosmologies). Each ten-dimensional brane supports a gauge field that expands and cools upon the brane, promoting condensation of stable matter and the emergence of complex structure. Robed in the dimensionalities of the branes, expansion continues over aeons, eventually dissolving matter and field into nearly silent quiescence. Finally, the separation between the branes collapses conformally, and the branes are reunited in restored supersymmetry, encompassing the rays of quantum action that are the seeds of rebirth for the next cosmic cycle.

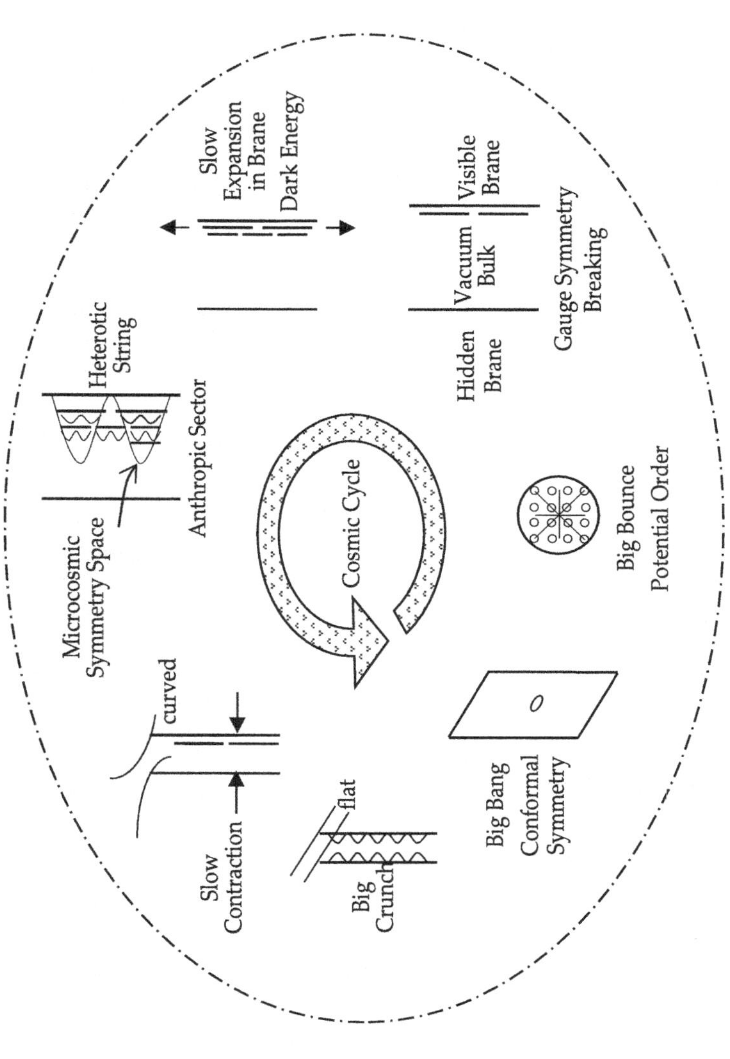

Figure 15-1 *Macrocosmic Symmetry Space*
Cycle of Conformal Braneworld Universe

The braneworld that emerged from the big bang consisted of a hot gas, made up of closed branes "wrapped" around internal-dimensional manifolds. [265] The manifolds were formed of compact loops (cycles) and tori of various dimensionalities, the prime example being the six-dimensional, 3-torus, Calabi-Yau manifold, which parametrizes the internal symmetry space of the standard model. The number of times a brane wraps around a dimensional cycle specifies the quantum number of its momentum, while the tension accruing from the curvature of a brane corresponds to its mass and charge. The current state of our universe is parametrized not only by the six-compact-dimensional manifold underlying the internal symmetry space of the standard model, but also the four-extended-dimensional manifold of relativistic spacetime, which emerged from the compact state.

The four dimensions of spacetime could only have become extended from their initially compact state if their wrapping and constraining branes and antibranes annihilated one another, freeing the dimensional manifold to expand to macrocosmic proportions. The Brandenberger-Vafa conjecture[266] proposes that the number of dimensions occupied by a manifold's wrapping brane determines the probability it will meet and mutually annihilate a symmetrically opposite, charge-conjugate anti-brane. Calculations show that the lower the dimensionality of a wrapping brane, the higher the probability it will encounter an annihilation intersection, and that the largest spatial dimensionality in which conjugate branes are likely to intersect, and annihilate, is three. Therefore, the number of un-compacted, extended spatial dimensions became three, as now observed, while the other six dimensions remained wrapped and compact.[267] That compact dimensionalities, of themselves, are dynamic and possess the potential energy to spring into extended form, is reminiscent of relativity and the framed standard model field theories with respect to the existence of causal relationships between dimensionalities and their component values.

Once the Brandenberger-Vafa process was completed, the residual six-dimensional internal symmetry space representations (particle types) had been sequestered onto the two individual 3-branes.[268] The external symmetry space was now comprised of *five* dimensions: the three spatial dimensions of the branes, the dimension of time, and a "new" dimension, interconnecting the two branes. Each of the 3-branes carried one member of a superpartner pair (Chapter 14). The aptly named "visible brane" is the one in which we live, and includes all of the particles in the cosmos that we can observe by virtue of the light they emit. In contrast, we cannot see light from the "hidden brane," because its initial complement of only heavy superpartners quickly degenerated into the lightest, but plentiful, supersymmetric particle, the neutralino, which does not participate in the photon-transmitting electromagnetic interaction.

The existence of invisible dark matter constituting the hidden brane has been inferred from the astrophysical dynamics of visible stellar bodies, whose motion, to be consistent with the laws of gravity, implies more matter than can be accounted for via observed photons. Quantitatively, the current distribution of energy in the universe is 74% for dark energy, 21% for the dark matter comprising the hidden brane, and only 5% for all of the luminous objects we see in the cosmos.

The visible and hidden branes are portrayed in Figure 13-1 as being flat boundaries of the included bulk, but in actuality the non-trivial geometric shapes, in addition to the compositions of both of these branes, as well as the bulk, are vital to their physics. Significantly, the negative spatial curvature of the bulk reflects the visible brane as a holographic dual. The Anti-deSitter Space/Conformal Field Theory (AdS/CFT) correspondence conjecture of Juan Maldacena says that our physical universe may be equally well described by holographically dual theories; a description of our universe in terms of a 5-dimensional (4-space, 1-time) bulk with only gravity, is equivalent to another description in terms of a 4-dimensional (3, 1) conformal field theory without gravity (the standard model) on the boundary brane. In the late phase of a cosmic cycle, when the hidden and visible branes draw together in a big crunch, global supersymmetry is restored as the branes unify.[269] Also in the big crunch phase, as the separation between the branes diminishes, local field curvatures (excluding Weyl conformal gravitational curvature) reach intensities that cause the branes' long wavelength forms to thermalize as quantum fluctuations, which are destined to spin-up in the next cosmic cycle in accordance with the covariant Lagrangian of the *pure spinor formalism.*[270]

The Lagrangian resulting from a big bang transformation includes both nonlocal and local terms (Chapter 15). The physics corresponding to the nonlocal terms is purely phasal (cyclic) and independent of spatial location; that is, an action occurring at some particular location may have an instantaneous effect at some distant location. Initially, the entire quantum universe was nonlocal, accommodating only topological and phasal relationships. In the ontological interpretation of quantum field theory derived by David Bohm, phasal relationships are expressed in terms of a bosonic wave-functional (a function of functions), which he named the *super quantum potential.*

The super quantum potential is the initial nonlocal physical influence that sets the parameters by which all of manifestation may evolve in the forthcoming cosmic cycle. In compliance with Curie's principle, broken symmetries are a necessary ingredient for this manifestation. Through a cosmic cycle, the effects of the local terms in the Lagrangian arise gradually through stages of spontaneous symmetry breaking, with the super quantum potential giving way to a succession of condensed quantum potentials. The breaking of the unstable (tachyon) symmetry of the initial super quantum potential initiated a cascade of brane-antibrane (matter-antimatter) condensations and annihilations.[271] When the temperature dropped below the pair production threshold, a small excess (one part per billion, due to an effect of charge-parity symmetry breaking) of ordinary matter remained, establishing the platform upon which our material world has been constructed.

The super quantum potential has an informational aspect resulting from its inherited homogeneity and its proclivity for symmetry breaking and integration. The topic of active information is discussed in Chapters 13 and 17, but let us note here that the super quantum potential is introduced into a cosmic cycle by a big bang transformation. Entering into a big bang, the afferent cosmic channel conducts the ultra-long wavelength residual form of the just-expended cycle into the big bang process. Emerging through the big bang event horizon, the efferent cosmic channel delivers the super quantum potential, a conformally re-dimensioned topological form imbued with a short-wavelength micro-structure (energy potential), homogeneity (information potential), and a heritable transformational nature able to instill the upcoming cycle with the potentiality for interactions and complex interdependencies.

The transformational character of the big bang is reiterated in the myriad interactions transpiring in the course of a cosmic cycle, presiding over all structures and events that come to pass. The concept of transformation has also played a prominent role in metaphysics throughout history. The fundamentality of the transformation was recognized by Greek philosophers, such as Heraclitus, when he proclaimed in the *Panta rhei,* that "Everything flows," and "Everything changes and nothing remains still."

Today, the transformation remains a basic principle of natural science. The Lagrangian of the standard model, for example, is cast in terms of the internal symmetry transformation groups, and that of general relativity in terms of diffeomorphism invariant transformations. Entities such as strings and branes are pictured as the basic, irreducible physical objects of M-Theory (as are particles in field theories) for the reason that "their" discrete charges are conserved through all transformations. But it is because these charges are conserved that symmetry transformations must take place, as proven in the "deep theorems" of the renowned algebraist, Emmy Noether. [272]

Emmy Noether had been asked to clarify the "improper" local energy conservation inconsistencies afflicting the newly formulated theory of general relativity. She was not permitted to personally present her work to the elite physicists of the day, who had assembled at the University of Gottingen to discuss the problem, because she was a woman. But Emmy Noether's profoundly important work would ultimately prove to be instrumental in the great discoveries involving symmetry and gauge theory throughout the 20th century.

> *Yesterday I received from Fraulein Noether a very interesting paper on invariant forms. I am impressed that one can comprehend these matters from so general a viewpoint. It would not have done the old guard at Gottingen any harm had they picked up a thing or two from her.*
>
> — Correspondence from Albert Einstein to David Hilbert [273]

The general emphasis of Noether's Theorems, focusing on symmetry transformations rather than concretized objects, was a step forward in natural science. In the context of macrocosmic cycles, even time and space arise as particular representations of symmetry group transformations, and are thus byproducts of a deeper, transformational *ontic structuralism.* It is to the transformational structures themselves, such as the U(1) group of electromagnetism, and not to their interchangeable representations, such as quark and electron, that ontic realism should be assigned.

> *... it is the group structure - as mathematically expressed in terms of the invariance under U(1) of the full dynamics represented by the Lagrangian - which is the holder of the physical meaning... structural realists do not commit themselves to an ontology of spatiotemporally fixed and object-like entities, such that worries about the empirical underdetermination of the*

object content of our best theories as well as about discontinuous shifts in theoretical entities in the course of scientific progress do not apply. What does, however, remain invariant or shows much more historical continuity is the structural content of scientific theories.

— Holger Lyre [274]

The ontic-structuralist conception of a transformational fabric of reality holds well with regard to current theories of cosmic evolution. While the concreteness of objects progressively loses substance as they are resolved back into their homogeneous origin at the big bang, the principle of the symmetry transformation remains. But ontologically real transformations define pathways, and these pathways form cycles from which levels of informational interdependency and self-organized complexity emerge. We are led to the model of a dynamic macrocosm embodied in a holistic web of coupled and nested cycles, holistically mandated to evolve and involve together in the expression of manifestation.

* * *

16

Cycles within Cycles

Cycles are ubiquitous in our universe. They are found in abundance in the symmetry space of the physical macrocosm as well as the microcosms of life. Cycles may form closed systems such as the earth revolving around the sun, or open systems such as living microcosms reciprocating with their environment. Quantum fluctuations in the fabric of spacetime, holistic autocatalytic sets in deep-sea vents, neural pathways in the human brain, and oscillating ten dimensional branes are all cycles. Cycles span all scales. In the macrocosm, the braneworld completes its evolutionary cycle once every trillion years or so, while in the small radius limit of M-Theory, closed string excitations circumnavigate toroidal manifolds in cycles of only 10^{-43} second.

A cycle is usually thought of as a temporally changing sequence of states that, upon completion of the sequence, attains a final state identical to the initial one. This conception of a cycle is acceptable in the classical regime, but in certain macroscopic and microscopic limits, where the separateness of space and time is transcended, the idea of the state of an object or a process pathway transforming through an immutable dimension of time must be superseded:

In the macroscopic limit of special relativity:

> *Henceforth, space by itself, and time by itself, are doomed to fade away into mere shadows, and only a kind of union of the two will preserve an independent reality.*
>
> — Hermann Minkowski [275]

And, in the microscopic world of quanta:

> *One can equally well say that a particle moves on a closed loop in a given fixed background, or that the particle stays fixed and space and time fluctuate around it.*
>
> — Stephen Hawking [276]

Even in these extreme contexts, where space and time are resolved into their underlying symmetry, the transformational ontic-structural cycle endures, exposing the fundamental foundation of physical reality. The theory of *loop quantum gravity* provides an insightful platform upon which the fundamentality of cyclic transformations makes contact with the emergent characteristics of the four-external-dimensional fabric of spacetime. Moreover, the loop quantum gravity model captures the cyclic properties of the microscopic domain in a formalism that is extensible to its macrocosmic counterpart, *loop quantum cosmology*,[277] which describes the transition through the big bang in terms of braneworld cycles. [278]

According to loop quantum gravity, the *holonomy* is the fundamental microscopic cyclic transformation. A holonomy is the limiting trajectory around a circular path as the diameter of the path approaches zero. The holonomic path, however, does not necessarily terminate at its starting point, and the measure of the deviation between the starting and ending points is the connection (discussed in Chapter 13) which specifies the local curvature of spacetime. As the name implies, holonomies have a holistic aspect: the phasal integration of holonomies constructs a gauge field. Loop quantum gravity is ontic structural in its premise that the cyclic holonomic loop transformation is the fundamental microscopic entity in the physical universe.

The cyclic transformations of loop quantum gravity produce discrete, minimum-sized slices of spatial area equal to 10^{-66} cm^2. The theory depicts these space-producing transformations as a "spin network" [279] of lines, or edges, whose ends meet at nodes. Edges connect the slices of space they represent into chunks of volume, which are represented by trivalent nodes, or, in the relativistic context of four-dimensional spacetime, by the tetravalent nodes of a "spin foam." The dimensions of a spin network of transformations are inverse to those of the space they generate. In this way, from the holonomic transformations that define the edges of a spin network, the dual format of curved space is generated, complete with stress, torsion, and black holes. The edges of a spin network further interconnect to form polymer-like transformation-structures, which fold and twist into "knots, links, and kinks," embroidering the texture of space with a dynamic collaboration of coupled and nested cycles.[280]

The internal symmetry space of the standard model is based upon cyclic dimensionality. Each of the six circular dimensions parametrizing the internal manifold supports a holonomic transformation whose particular representations define the gauge properties of a particle species and its interactions in the external symmetry space. We have seen that new levels of interaction emerge in the external symmetry spaces on the basis of these gauge interactions, able to support cyclic process-structures, self-organization (entropic and selective), and complexity. Indeed, in the anthropic sector, the integration of coupled and nested cycles in the electrochemical regime has risen to the high degree of complexity that characterizes the neural processes of the brain.

All cycles and their complexes form *objects* that interact and integrate in conformance with an information-based paradigm which pervades nature. Objects possess the intrinsic properties of methods (library of what they can do) and hierarchical nesting, as well as the extrinsic capabilities of messaging, coupling, and inheritance. For readers interested in objects, on the history and principles of object orientation as applied in computer science, and some important examples of objects found in nature, are provided in Chapter 10.

We observe that from the primitive seeds of coupled (linked) and nested (embedded) cycles in the spin networks of loop quantum gravity, the elaborate systems of life have blossomed. The variety of cycles underlying microcosmic organic life and its reciprocative macrocosmic environment spans extensive spatial and temporal scales. For example: spinning molecular machines, electrochemical autocatalytic cycles, pacemaker cells synchronously firing on repetitive schedule, biological clocks linked to the daily light-dark circadian rhythm, the 28 day menstrual cycle of the human female, the reproductive period of the seventeen year cicada, and the inter-twining thalamocortical pathways of the brain, are all cycles in the life systems of earth.

All life forms store a type of potential energy necessary to uphold their constitutional and functional integrity. This energy is drawn by, and metabolized for, the orderly networks of coupled and nested cycles comprising self-proficient organisms; see Figure 16-1a.[281] The measure of ordered energy usage is called *ectropy* (or negative entropy). Ectropy is the opposite of *entropy* (S). Entropy measures the dispersal, or dissipation of energy, whereas ectropy refers to coordinated, interdependent energy transformations. In organized living systems, cycles have high ectropy and low entropy ($\Sigma\Delta S = 0$); see Figure 16-1b.[282]

Energy, matter, and information flowing into an organic system distinguish the types of signs that complement its process-structure. At the ontic-structural level, however, matter, energy, and information are equivalent with respect to the microcosmic symmetry space. Ontic-structural symmetry in the microcosm is a fundamental principle in the macrocosm as well. The efforts of Claude Shannon, "the father of information theory," to quantify the amount of information contained in, for example, an email message, led him to discover, unexpectedly, a formula with the same form that was previously derived by Ludwig Boltzmann to quantify thermodynamic entropy.

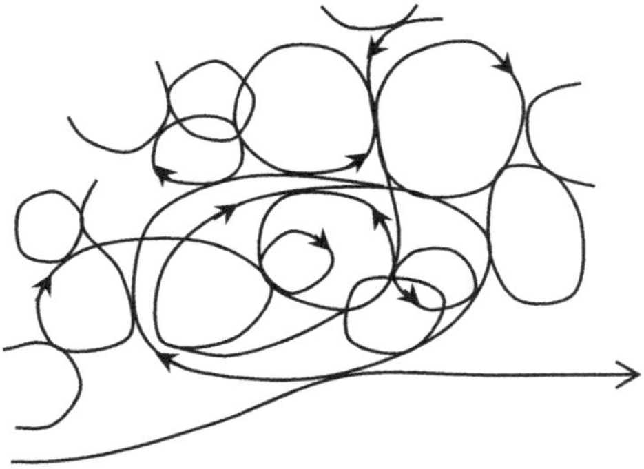

a. The life-cycle of an organism consists of innumerable, nested and coupled sub-cycles within.

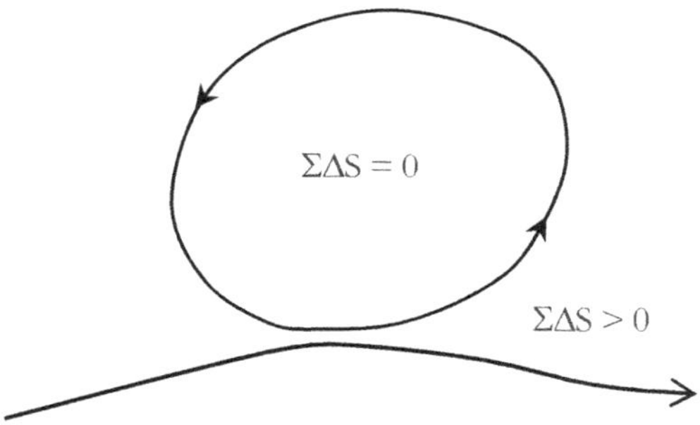

b. An organism consists of cyclic non-dissipative processes coupled to the dissipative one-way energy flow.

Figure 16-1 *Cycles within Cycles*

Boltzmann entropy is quantified in units of energy divided by temperature, while Shannon entropy is expressed in bits of information. The differences between thermodynamic entropy and information entropy are merely in the conventions of the discipline.

> *Thermodynamic entropy [energy and matter] and Shannon entropy [information] are conceptually equivalent: the number of arrangements that are counted by Boltzmann entropy reflects the amount of Shannon information one would need to implement any particular arrangement...*
>
> — Jacob Bekenstein [283]

High entropy refers to a state in which potential degrees of freedom for a system are maximized over all space-time domains. Statistically, this is the *equilibrium attractor* condition to which *isolated systems* (isolated from an environment) tend to settle over time—in accordance with the second law of thermodynamics. However, *open systems,* which do exchange energy, matter, and information with an environment, may attain a state of high ectropy in which activities in its temporal and spatial domains tend toward ordered coupling; an *attractor of complexity* that is far from the equilibrium attractor. Self-organization is the inevitable consequence of the laws of far-from-equilibrium physics.[284]

The biosphere of earth is a *far-from-thermodynamic-equilibrium* system that operates on the temperature gradient between the sun, the system's heat source, and outer space, the system's heat sink. This temperature gradient enables potentially usable energy and ectropy of sunlight to be utilized in the development of ordered systems on earth. Individual photons emitted by the hot sun have high frequency, thus high energy; those absorbed on earth and reemitted back into cold space have lower energy. Since the amount of solar energy received on earth must be equal to the amount returned to space—as required by the "first law of thermodynamics," or conservation of energy principle—it follows that the number of individual photons (quantum bits of informational data) emitted by earth must be greater than the number received. The photons which are thereby *captured* by earth's biosphere, through photosynthesis, provide both the energy and ectropy necessary for the binding and ordering of molecular structures and processes of life. The flow of ectropy into the earth's biosphere via photosynthesis, facilitated by the gradient between the hot sun and cold space, provides the means by which increase-ingly complex process-structures evolve into stable organic systems that are *far-from-chemical-equilibrium.*

In addition to photosynthesis, there are a number of mechanisms that harness ectropy from far-from-thermodynamic-equilibrium gradients and organize far-from-chemical-equilibrium life systems. The original mechanism in effect on earth four billion years ago utilized the heat source of the earth's molten core to promote the growth of microbial life on the surface of the hydrothermal deep-sea vents protruding from the mantle into the heat sink of the enveloping, colder ocean waters. Many of these forms of life still flourish. Additionally, certain organisms, called chemotrophs, obtain their energy and ectropy by oxidizing inorganic chemicals, or by eating other ectropy-bearing organisms further down the food chain.

Far-from-equilibrium systems are open systems in that a thermal or chemical flow through the system is required to support and sustain its behavior. Intrinsically, open systems consist of internalized cycles. As noted in Chapter 4, a key theorem of biochemical thermodynamics proven by Harold Morowitz sheds light on the cyclic nature of open systems: "The flow of energy through an open steady state system will lead to at least one cycle within that system."

Self-organization is thus the consequence of cyclic, far-from-equilibrium physical laws. The evolution from the disorder of chemical equilibrium to the order and complexity of cyclic process cycles typically occurs in phases of symmetry breaking, called phase transitions, each involving a nonlinear thermodynamic Lagrangian. Complexity can only arise in an environment where order and disorder coexist. A completely ordered system such as a crystalline structure, or a completely disordered system such as a gas in thermal equilibrium, lack complexity; that is, integrated causal relationships.

An organism is a steady state system, open to its environment for the flow-through and processing of matter, energy, and information, thus making the organism biosemiotic. The biosemiotic relationship between an organism and its environment is triadic and includes the process-structure of the organism as interpretant, with built-in afferent and efferent channels evolved to accommodate the optimal flow-through of signs. In the case of a human being, matter processing brings about biosynthesis, energy processing sustains homeostatic metabolism, and information processing takes biogenetic, neuronal, semiotic, and other forms. The largest-scale afferent and efferent channels serving human processes include the digestive/excretory system for matter flow-through, the respiratory O_2/CO system for energy flow-through, and the nervous system for sensory-limbic/executive-motor information flow-through.

True to the transformational, ontic-structural paradigm, the flow-through principle is fundamental. With regard to matter transfer, the essentiality of afferent and efferent biosemiotic channels to the process-structure of a human interpretant is demonstrated by the fact that while the overall steady state of the human process-structure may function for many decades, every particle of matter that constitutes its corporeal body is temporary, flowing through its afferent and efferent channels within a few years, thus resulting in a life-series of completely different corporeal bodies (albeit a series of bodies that develops through stages from embryo to adult—in accordance with the informational content of its matter-replenished, afferent-efferent, open system DNA processor).

The functioning of both abiotic and biotic open systems involves process cycles that are either: confined within, passing though, or outside the boundary separating the system from the environment. In Chapter 3, we saw that the flow of energy, or chemicals, through an open system can lead to self-organization within the system, as in the Raleigh-Bernard convection dish and the Brusselator reactor experiments.[285] In the convection dish experiment, certain chemical mixtures in a shallow dish (the open system) self-organize into convection cells when heated from below (energy flowing through the open system), to form a hexagonal, spatially ordered, global pattern; whereas in the Brusselator experiment, chemical mixtures self-organize temporally ordered fluctuations of color.

There are two primary factors that distinguish self-organizing abiotic open systems such as the convection dish and the Brusselator, from the biotic systems of life: the enormous degree of complexity among the sub-cycles within the boundary of a living system, and the holistic closure of these cycles—as distinct from, but conjugably connected to, the cycles of the environment outside the boundary. All self-organizing biotic systems interact with their environments by means of three types of cycles:

> C1 cycles are executed and completed internally to the horizon boundary of an organism.
>
> C2 cycles are executed and completed through the horizon boundary.
>
> C3 cycles are executed and completed externally to the horizon boundary.

C1 cycles are executed internally to the horizon boundary of an organism. As its name implies, the cardiovascular "circulatory" system of a human being represents a classic example of a C1 cycle: oxygen-rich blood issues from the heart's left ventricle, absorbs nutrients while flowing via arteries into the interstitial fluid, diffuses its oxygen-nutrient content into cells (nested cyclic processes), picks up carbon dioxide-waste from the cells, flows via veins to the right atrium, and back into the heart—having coupled with digestive, respiratory, and endocrine process cycles along the way.

C1 cycles are abundant in human beings: biochemical pathways, neurotransmitter reuptake, ATP hydrolysis, etc. Short-circuit crossover links between afferent and efferent channels also fall into the category of C1 cycles, as exemplified by the familiar *feedforward* patellar (knee-jerk) reflex action, which bypasses the brain via a direct connection between the afferent sensory and efferent motor channels of the nervous system. In contrast, thinking in words (described in Chapter 7) involves a *feedback* nervous system crossover connection from the efferent channel of verbal expression emanating from the Broca area of the brain (which sequences language elements for output), to the afferent auditory channel, into the Wernicke area (responsible for language comprehension).

C3 cycles are executed externally to the horizon boundary of an organism. C3 cycles do not couple to an organism's internal C1 cycles by direct transfer of energy, matter, or information through the horizon boundary. However, environmental C3 cycles may couple to the internal C1 cycles of an organism indirectly, via C2 cycles passing through the boundary; there may be afferent transfer of energy, matter, or information from the environment (C3), across the boundary (C2), and into the organism (C1), as well as efferent transfer of product, if only entropic heat energy, by the reverse pathway.

The nitrogen cycle, critical to life on earth, is a straightforward process in which atmospheric abiotic nitrogen settles into the soil of the ground (C3), is assimilated (C2) by an organism, processed (C1) by the organism, and returned (C2) to the atmosphere (C3), having completed the cycle by returning to the initial state.

A more abstract network of cycles, involving the phenomenal symmetry space of a human being opening an umbrella in response to the detection of rain, is illustrated in Figure 16-2.

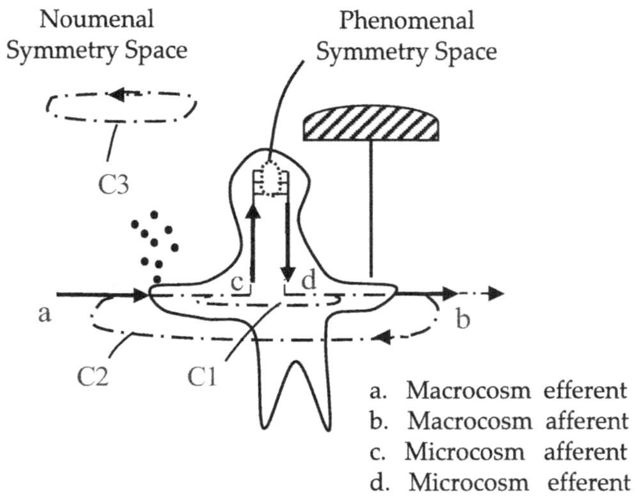

Figure 16-2 *Rain!*

The types of cycles involved in this example include a C3 cycle (the formation of rain in the environment, the "water cycle"), a C2 cycle (sensory transfer through the horizon boundary into the microcosm), a C1 cycle (cognitive processing within the internal symmetry space of the microcosm), and the biosemiotic completion of the C2 cycle (motor transfer through the microcosm's horizon boundary into the external environment). Specifically, rain in the environment reacts with appropriately structured receptor cells on the surface of the hand (a). An afferent electrochemical process pathway then supplements and relays signals generated by these reactions to the focused symmetry space of the brain (c). Afferent neurons in the brain fire their axons synchronously and collectively across synaptic clefts when sufficiently induced by the feedback potentials (apprehension) of efferent channel dendrites. The firing produces an informational pattern which forms an object of consciousness: "It's raining." The interpretant's experience of the translated environmental condition is followed by the firing of efferent neurons (intention), producing a cascade of motor signals (d) that guides the appropriate muscles to open the umbrella (b).

In every cosmic epoch, structures and cycles are unified in potentiality at the big bang, then manifested in the building of complexity and life, and dissipated under the unrelenting entropic force. The entropic force is a macroscopic force that acts upon a system in a way *not* determined by the properties of microscopic forces (such as electromagnetism), but by the whole system's statistical tendency to increase its entropy. For the purposes of a developed organism somewhere on the anthropic pathway, the persistent tendency for its order and complexity to dissipate to a state of equilibrium must be overcome.

> *It is by avoiding the rapid decay into the inert state of 'equilibrium' that an organism appears so enigmatic... What an organism feeds upon is negative entropy [ectropy], or, to put it less paradoxically, the essential thing in metabolism is that the organism succeeds in freeing itself from all the entropy it cannot help producing while alive.*
>
> — Erwin Schrödinger [286]

An organism requires replenishment of its ectropy, which, as described above, is related to the flow of signs (matter, energy, ectropy) through its open system, necessarily cyclic, process-structure. The complexity of the organism depends on the degree of interdependency among the cycles unifying its process-structure. In a changing and volatile environment, the precise and optimal coordination of the development, activation, inhibition, and fortification of the cycles entangled in the complex maze is a daunting task. From the early days in deep-sea vents, catalytic cycles and their descendants have met the challenge.

A *catalyst* is a substance that lowers the activation energy for a particular reaction, thus increasing the rate of the reaction, without itself being consumed by the reaction. When there is an adequate source of reactant supplying the reaction substrate, the action of a catalyst on that reactant will increase the amount of the given product. This increase is effected by the emergence of new cycles, in response to the presence of the catalyst, intermediate to the substrate and the products. Catalytic process cycles can be effective; for example, one ounce of the catalyst crystalline pepsin can produce more than one ton of product per hour.

The affinity of a molecule for binding a substrate reactant, including a catalyst, varies with the shape of the electromagnetic fields enveloping material "ball and stick" structures. The binding of a reactant to, and the dissociation of product from, a catalytic molecular complex occurs at its active and allosteric ("other") sites, respectively. These sites are generally protrusions or depressions in the overall shape of the catalyst's field. Binding of a molecule at its active site causes the shape of a catalytic complex to change at its allosteric site, thereby influencing subsequent interactions with the substrate. This influence on the substrate may only be an electromagnetic ripple, but stable hardwired sub-cycles may still develop. In biological systems, allosteric regulation of catalytic cycles is often further enhanced by the binding of an effector molecule at the catalyst's allosteric site, providing an additional boost or inhibition to the activation potential. Allosteric regulations are an example of advanced control-loop tuning employed by nature, improving both loop-back from downstream products and feedforward to upstream substrates.

It may happen that a product of a catalytic cycle is a copy of that cycle. This is the process of *autocatalysis*, which has far-reaching consequences common to all life forms. The autocatalytic capabilities of RNA and DNA molecules, for example, is the basis upon which an organism implements its self-replication and self-sustainability. An *autocatalytic set*[287] is a collection of molecules that catalyze each other's production. An autocatalytic set may comprise a simple, closed-loop hypercycle (which is a sequence of reactions in which each molecule in the cycle exclusively catalyzes the production of its successor, such that the last reaction catalyzes the first), or it may comprise an intricate network of coupled and nested catalytic pathways. Autocatalytic sets act collectively, forming a self-sustaining, self-organized whole that can reconstitute itself after a large portion of the set has been removed. An autocatalytic set is depicted in Figure 4-2.

The equations that describe autocatalytic sets are non-linear. The rate of production of a product (including catalysts) is proportional to the square of its concentration, so that the processes with the highest reaction rate coefficients for this product quickly dominate its consumption, thereby implementing the process of "chemical natural selection." [288] At this local level of molecular interaction, natural selection appears as a competitive, selfish,[289] survival of the fittest process that takes place between process cycles vying for limited resources. But in the holistic context of autocatalytic sets, encompassing their offspring forms of life, cycles implement cooperative, altruistic, symbiotic interrelationships as well.

We have discussed how the principle of cooperation applies outside the C1 cycles of a holistic autocatalytic set; for example, when individual cells export their own fitness to merge into a multicellular organism, when C1 organisms and their C3 biosystems mutually catalyze through the C2 cycles of Gaia, when neural complexes synchronize to produce integrated information, when organisms unite to establish societies, and so forth. The principles of coming together, of intertwining cycles cooperating at all scales, apply throughout all the provinces of biotic systems and their reciprocative environments. The insightful quote attributed to Margulis and Sagan in Chapter 4, in the context of multicellular emergence, now bears reiteration as a general principle, for the basis of human evolution lies in this law of nature: "Life did not take over the globe by combat, but by networking." [290]

Life is a microcosm of cycles, enfolded within a macrocosm of cycles, unifying the one and the many in a holistic network of ascending complexity. The seven stages of evolution on the anthropic pathway have emerged by way of cycles; from the cycles of string quanta, to the cycles of the brain, to the cycles of the cosmos.

*   *   *

17

The One and the Many

A big bang is a conformal transformation that restructures the symmetry space of the universe. The dimensionalities that parametrize the cosmic symmetry space and the data which quantify them transform in character as the universe transitions from the end of a cosmic cycle, through a big bang event, into the super quantum potential of active information that initiates the next cycle. Since space and time are dimensionally supersymmetrized immediately following a big bang, the transformation from one cosmic cycle to the next should be conceived as a horizon boundary associating holographically related symmetry spaces. From our location in the current, symmetry-broken cosmic cycle, however, the super quantum potential (Chapters 15 & 18) appears to be the initial chronological phase of cosmogenesis, preceded only by the big bang event (Table 14-1).

The super quantum potential impels the dissemination of the active information that is inherent in the low-entropy of the newborn universe's gravitational homogeneity and isotropy. This makes the assemblage of orderly structures and processes possible, carried out through symmetry breaking, nonlinear gauge effects, far-from-equilibrium processes, and emergence. Active information is the subtlest form of information in the macrocosmic symmetry space, enfolded in the potential of the "implicate order." [291] It is nonlocal, apparently enabling instantaneous actions at a distance, but in actually the implicate order transcends the relatively lower-dimensional reference frames unfolding in the classical, local "explicate order." In David Bohm's terms, the implicate and explicate orders share a cyclic relationship that unifies the "enfolded" whole with the "unfolded" parts. Thus, the *many* separable, local elementary particles of the explicate order are recursive cycles, unfolding from and enfolding into the *one* underlying unity of the "holomovement." [292]

The ultimate fate of the super quantum potential is to expend its active information in the holomovement by seeking the extremum implicit in its selective Lagrangian; specifically, the attractor of complexity that is realized in the form of a conscious microcosm. The mechanism by which the super quantum potential unfolds its information is phasal, and its constructions in the explicate order are imbued with the phasal seeds of cyclic interrelationships. Phasal synchronicity and harmony manifest in numerous scales and compositions. In the human brain, where explicate unfoldment is highly evolved, the synchronous, resonant, and phase-locked neural cycles of the cerebral process-structure integrate objects of consciousness. Active information is the holistic expression that *informs* manifestation how to select for and harmoniously construct microcosms, which ultimately and inevitably realize consciousness.

As the super quantum potential releases itself through the symmetry breaking hierarchies of the holomovement, certain threshold expectation values are attained which permit quantum fluctuation cycles to stabilize locally, differentiate with respect to global condensates, and integrate. These standing wave, sub-whole solitons (solitary self-reinforcing wave packets sustaining a localized conformation) effectively populate the explicate order, obeying their own reduced symmetry space Lagrangians, or *quantum potentials.* The implicate order determines which sub-wholes become actualized in the explicate order, as well as the nature of their relatively independent and stable forms. For example, an elementary particle is a sequence of sub-whole waves, outgoing and ingoing between the implicate and explicate orders, with successive waves in proximity to one other.

> *The super-quantum potential depends on the state of the whole universe and not only determines and organizes its sub-wholes, but also gives form to what has until now been called the elementary particles out of which everything is supposed to be constituted.*
>
> — David Bohm

Bohm's ontological interpretation of quantum theory "emphasizes the wholeness that is signified by quantum laws." [293] In this view, there exists a master wave-functional governing the entire universe. The laws of quantum mechanics, such as the Schrodinger equation (conservation of quantum energy), describe how wavefunctions evolve over time. Bohm rewrote the Schrodinger equation in polar coordinates to separate it into classical and quantal terms. Decomposition revealed the classical term as a function of position but not of phase, and the quantal term as a function of phase but not of position. Thus, the quantal term, the quantum potential, is nonlocal and holistic. In a historical twist, Bohm wrote a book entitled *Quantum Theory,*[294] explicating the orthodox interpretation of quan-tum mechanics. Einstein offered high praise, declaring it the clearest exposition of the quantum formalism he had ever read. Shortly thereafter, however, Bohm rejected the orthodox interpretation, but not the formal-ism, in favor of his holistic ontological interpretation.

Lower-dimensional sub-wholes can be either bosonic forms or fermionic particles. It is natural for sub-wholes to collectively synchronize and polarize their phases. The principle of stable collective synchronization was formulated in the 17th century by Christian Huygens, after he observed that pendulum clocks, which were his own invention, always synchronized when hung on the same wall. Collective synchronization is commonplace in the great assortment of harmonic oscillators found in nature, including atomic interactions, pacemaker cells in the heart, neural networks in the brain, and a chorus of crickets chirping their serenade.

An extension to general relativity theory called semiclassical gravity treats space as *one* continuous field, while energy is quantized into *many* discrete chunks (in this respect, semiclassical gravity differs from loop quantum gravity and string theory, wherein both space and energy are quantized). In semiclassical gravity, the curvature of space, G(x), is treated as a function on a continuous field of spatial points, x; while the distribution of matter throughout space, T(Ψ), is described in terms of a wavefunction, Ψ(x). Ψ(x) is a field whose value *squared* equals the *probability* that a discrete quantum event will occur at a particular point in the spatial continuum.[xv] The concept of a continuous spatial manifold, upon which to represent probability distributions for discrete particles, is generally adopted in the orthodox interpretations of quantum mechanics. This hybrid model is accepted pragmatically due to its success predicting experimental results. Unfortunately, epistemological paradoxes continue to plague these conventional interpretations: the source of uncertainty, wave-particle duality, nonlocality, etc.

In the orthodox interpretation, the phasal, nonlocal, non-classical term in the Schrodinger equation (the imaginary part of the Schrodinger equation expressed in polar coordinates) is taken to represent uncertainty inherent in the *present* state of a physical system. Uncertainty in the ontological interpretation is associated with the *past,* reflecting the impossibility (the minimal, but unavoidable uncertainty) of knowing precisely all of the initial conditions required to calculate the present state. In the ontological interpretation, this phasal term is elevated to be the *physically real* quantum potential, *Q,* (of the implicate order), which has profound causal consequences. (The symbol *Q* is employed below to signify the class of functionals that encompasses the "super quantum potential" as well as its descendant, symmetry-broken, "quantum potential" factors.)

Q is not subject to the local dynamics and configurations of the classical symmetry space. *Q* is a nonlocal term in the Lagrangian (the Schrodinger equation), which effectively embodies a holistic "guiding wavefunction," the *one* wavefunction for the *many* particles. The cosmic cycle unfolds through the expression of active information from its potentiality resident

---

xv The semiclassical field equation of general relativity is $G_{uv} = 8\pi G\,[T_{uv}]_\Psi / c^4$, where G_{uv} represents the curvature of *(continuous)* space and, equivalently, the gravitational field. T_{uv} is the energy distribution in terms of the positional expectation value $[\ldots]_\Psi$ of the *(discrete)* quantum state of the matter fields Ψ.

in the *form* of Q. The phasal form of Q induces the recursive cycling of sub-whole refreshment and dissolution, stimulating the production of the standing-wave, stationary state, solitonic "particles," described above. A particle subdivision of the wavefunction is never completely isolated, as it is always affected by Q. Based upon phasal signaling of active information by Q, the mass-energy of a local particle is repetitively drawn from its underlying substrate. In the case of a particle such as an electron, its own self-energy is derived from the global vacuua of virtual particles and condensates. Note that this model does not preclude the possibility that elementary particles—on Calabi-Yau manifolds, for instance—may possess an inner structure capable of processing active information semiotically.

> *I do not doubt that the day will come when that description too, will have to yield to another one, for reasons which at present we do not yet surmise. I believe that this process of deepening the theory has no limits.*
>
> — Albert Einstein

The wavefunction formalism is routinely applied to sub-whole systems to *approximate* their quantum behavior. In particular, when the composite wavefunction for a system consisting of two identical particles is considered, the counter-intuitive *exchange symmetry* of nature must enter into the Lagrangian. Exchange symmetry is derived from a fundamental postulate of quantum statistics which states that no observable physical quantity is altered by exchanging the locations of two identical particles. In classical physics there are two ways particles can be distinguished: by differences in their intrinsic physical properties such as mass and electric charge, or by differences in their spatial locations. Empirical evidence has shown that the laws of quantum mechanics differ from those of classical physics and thermodynamics in terms of the statistical probabilities of composite wavefunctions.

To illustrate quantum mechanical exchange symmetry, imagine spatially separated points, a and b, as tiny boxes; and two particles, p and q, that are identical with respect to their intrinsic properties. In classical physics, there are four equally probable arrangements for the particles p and q to be in boxes a and b: (p and q in a), (p and q in b), (p in a and q in b), (q in a and p in b)—per classical Maxwell-Boltzmann statistics, in which particles with identical intrinsic properties are spatially distinguishable. In the quantum world, however, with p and q still indistinguishable, two other statistical scenarios are observed. In the first quantal case, there are three equally probable arrangements: (p and q in a), (p and q in b), (? in a and ? in b)—per Bose-Einstein statistics (where ? reflects the uncertainty principle). In the second quantal case, there is only one possible arrangement: (? in a and ? in b)—per Fermi-Dirac statistics. Particles that interrelate, or superpose, in accordance with Bose-Einstein statistics are the *bosons;* and particles that superpose in accordance with Fermi-Dirac statistics are *fermions.* All standard model particles conform to these quantum statistics and behave non-classically, e.g., bosons such as photons and gluons, and fermions such as electrons and quarks.

Exchange symmetry, and consequently the properties of bosons versus fermions, derives from the phasal, complex-valued composition of the wavefunction, Ψ. Since the observable, measurable physical properties of particles depend on Ψ^2 (complex conjugation), the laws of quantum mechanics mathematically (and empirically) allow two possible ways to obtain an observable (Ψ^2); namely, $(+\Psi)^2$ and $(-\Psi)^2$. The composite wavefunction representing the symmetric possibility for two identical particles, $\Psi(a, b) = \Psi(b, a)$, describes the bosons, and the antisymmetric possibility, $\Psi(a, b) = -\Psi(b, a)$, describes the fermions (i.e., exchanging the locations of two identical fermions, a and b, changes the sign of their composite wavefunction).

Fermion exchange antisymmetry was first conjectured by Wolfgang Pauli to provide an explanation for the restricted numbers of electrons found to occupy the energy-level shells of atoms. Statistically, if two electrons approach one another (so that as a ~ b, then $\Psi(a, a) \sim -\Psi(a, a)$, implying $\Psi^2 = 0$), then the probability for two electrons to be coincident is $\Psi^2 = 0$. Therefore, two electrons coming arbitrarily close to one another is not mathematically possible. Fermions must keep their distance to obey one of the most important principles of quantum physics, the Pauli Exclusion Principle. This space occupying restriction on fermions underlies their role as the structural constituents of matter (e.g., atomic nuclei, atoms, molecules), keeping it from collapsing in upon itself. In contrast, identical bosons have a tendency to integrate, or clump together, because the probability for them to be proximate (1/3 per Bose statistics) is relatively greater than for the classical case (1/4 per Boltzmann statistics), where quantum mechanical effects are negligible. The space occupying behavior of bosons underlies their role as integrators of holism and symbiotic relationships; bosons are the constituents of radiation and the messengers of the gauge interactions; see Figure 17-1.

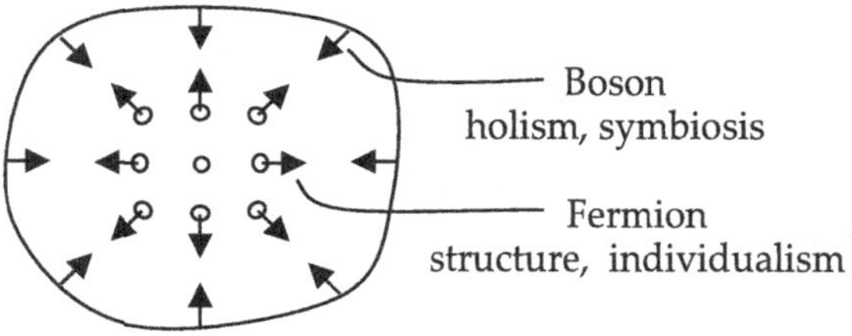

Figure 17-1 *Exchange Symmetry*

The One and the Many

In the orthodox (Copenhagen) interpretation of quantum mechanics, a fermion exhibits two alternating modes of behavior: coherence and decoherence. In the coherence mode, the fermion evolves like a complex-valued wavefunction; while in decoherence, the wave-packet collapses and displays classical properties. Thus, in the orthodox interpretation, an entity such as an electron is considered to alternately exhibit wave *or* particle behavior. The wave and particle modes are said to be in a relationship of complementarity, in which a sub-whole jumps back and forth between its wavefunction and classical identities, where different laws of physics apply. Historically, this interpretation of quantum behavior bred apparent paradoxes and much confusion, even for the founding Nobel laureates of quantum theory:

> *If all this dammed quantum jumping were really here to stay then I should be sorry I ever got involved with quantum theory.*
>
> — Erwin Schrodinger's lament to Neils Bohr [295]

Dilemmas such as indeterminism and nonseparability, which trouble the orthodox interpretation, vanish in the ontological interpretation, wherein the one holistic wavefunction *and* the many enfolded sub-whole particles of the lower (external) dimensionalities are simultaneously real. The ontological interpretation fits naturally with decoherence theory, the envariant framework in which sub-wholes and the environment are conjoined, as previously discussed in Chapter 3 in the context of quantum Darwinism. In decoherence theory, a coherent sub-whole, superposed in the universal wavefunction, "leaks" its (phasal) interference term into the environment, leaving a decoupled classical particle. Decoherence is an extremely sensitive process. The threshold sufficient for environmental mass-energy density to induce decoherence of a solitonic wave-packet is miniscule, requiring the mere presence of a speck of space dust, a single graviton, or even a microscopic black hole.

In the ontological interpretation of quantum mechanics, decoherence of a classical particle from the superposed wavefunction is equivalent to the generation of a particle by the quantum potential. The loss of the phasal interference term in decoherence theory is identical to the loss of the phasal quantum potential term in Bohmian mechanics. In the orthodox interpretation, the local mass-energy curvature in the explicate order induces recursive decoherence-recoherence cycles with period times of order 10^{-30} second. According to the ontological interpretation, coherent active information from the phasal, cyclic form of the quantum potential repetitively instructs a sub-whole to draw energy from the vacuum. (It appears that harnessing the coherent state of matter is technologically feasible. A current case in point is the engineering effort directed toward taming the propensity for decoherence imposed by our dense, dynamic terrestrial environment. Theoretically, a sustained superposition can be exploited by the nanotechnologies—semiconductors, intra-molecular chemistry simulations, etc.—and by higher-dimensional, massively parallel "quantum computers," which have been successfully implemented in the field of cryptology.)

Phase is a mathematical entity that refers to rotation or spin, as position refers to translation or velocity. The wavefunction has phase as a parameter. Elementary particles have quanta of spin, empirically detectable in the form of angular momentum and magnetic dipole moment. It had long been assumed that rotations in classical three-dimensional space could be described in terms of vectors—mathematical structures that specify both magnitude and direction. However, to describe quantum spin, decomposition of vector dimensionality into dual-channel *spinors* was found necessary to account for the non-commutative (asymmetric) behavior of fermions. The quantum state (the wavefunction sub-whole) of a particle's spin can be represented by a two-component spinor $\Psi = [\alpha, \beta]$, where α and β are complex numbers. These "two-valued quantum degrees of freedom," as the wavefunction spinor was described by Pauli, were later extended by Paul Dirac into compliance with relativistic spacetime, by use of a four-complex-component wavefunction spinor for fermions.

Spinors are basic elements of modern symmetry group theory, having transformational properties that are dependent on complex variables, which, by definition, involve the imaginary number *i* (the square root of negative 1), as well as even deeper roots of the number 1. The need to supplement the (observable) real number quantities with (unobservable) imaginary numbers was realized as early as the first century AD by Heron of Alexandria in his study of the frustum of a pyramid. Later, imaginary numbers were recognized to be necessary for solving such simple algebraic equations as $x^2 + 1 = 0$. Today, the long list of successful theoretical achievements based on the foundation of imaginary numbers assures that they, and their further decompositions, are the necessary and sufficient mathematical structures for describing the physical world at its deepest levels, including the unobservable implicate wavefunction and the observable explicate classical order it unfolds.

That unobservable is whole and perfect.
This observable is whole and perfect.
Because this whole has come out of that whole.

— Brhadaranyaka Upanisad

We have seen that the holistic wavefunction enforces oneness in the universe, while giving birth to the symmetry spaces of the many. As with the macrocosm, there is a holistic causal force in the microcosm that governs its evolution. This will be the subject of the next chapter, where we travel the anthropic pathway from the big bang to the place we call home, the planet earth.

* * *

18

The Anthropic Pathway

FROM THE BIG BANG TO LIFE ON EARTH

The evolution of the universe is a very powerful affair. Hypernovae exploding with the brilliance of 100 billion suns, exotic structures such as neutron stars rotating thousands of times per second, halos of invisible matter enveloping galaxies with supergiant black holes at their centers, and brane worlds oscillating, morphing, and interacting in the great cosmic rhythm; all features of a universe so potent that one would expect them to dominate the Lagrangian of the universe. Yet, there is a fragile, complexly-ordered branch flourishing on the tree of evolution, the one that bears the fruit of life, the one where we experience consciousness—the anthropic pathway. The anthropic pathway has relentlessly threaded its way along an extremum of the universal Lagrangian, pursuing an attractor of complexity where life and consciousness thrive.

The terms "anthropic pathway" and "anthropic sector" are introduced to draw a distinction from the term "anthropic principle," which is frequently employed in philosophies of physics and metaphysics.[296] The anthropic principle concerns any philosophy that deems consciousness to be an innate or predestined property of the universe. The "cosmological anthropic principle" and the associated problem of fine-tuning were discussed in Chapter 3. The cosmological anthropic principle is specific, addressing the question of how the constants of nature possess precisely those fine-tuned values required for complex conscious systems, such as ourselves, to exist.

The cosmological anthropic principle is invoked to compensate for the deficiency of physical theories to account for the fine-tuned constants of nature. Based on the M-Theory formalism, for example, there are many "landscapes," or possible universes, that *could* exist which are compatible with its Lagrangian—10^{500} universes per typical estimates, comprising a "multiverse." According to the logic of the cosmological anthropic principle (technically, a conjecture rather than a principle): if each of these multitudinous universes were to actually exist, then the probability that one of them would possess the fine-tuned values measured for our universe is reasonable; and since there does exist a universe having these fine-tuned values, namely ours, then all of the possible parallel universes *do* exist.

The cosmological anthropic principle is not economical, as it postulates an immense number of other universes in order to reconcile unexplained physical data about our own. While it is possible that other "universes" exist beyond the event horizon of our own, and that their existence could be inferred from characteristics of our own, all we could ever know about them (otherwise, they would *not* be other universes) is that they cannot support conscious life, because the premise of the conjecture is that our anthropic universe is unique, or at least exceptional, in that regard.

In contrast to the multiverse proposition, the notion that the process cycle of our holistic universe, illustrated in Figure 13-1, gives rise to a complex, self-consistent, open-system anthropic sector, explains the convergence of the constants necessary to support conscious microcosms. Whereas the cosmological anthropic principle excludes non-reductionist (top-down) guiding principles, the anthropic sector has evolved precisely because top-down symmetry principles together with an attractor of complexity have guided our universe along the anthropic pathway to its current level of complexity, to the closure of holistic living microcosms, and to vehicles of consciousness endowed with an endogenous, subjective experiential perspective.

The anthropic pathway traces an extremum in the macrocosm where its symmetries are maximally broken and where evolving electromagnetic process cycles self-organize toward maximum microcosmic complexity. The anthropic process self-organizes and fine-tunes toward an attractor of complexity (despite the enormous number of possibilities in the entropic phase space, and in lieu of the 10^{500} possible universes in M-theory) which is innate to the cosmological topology, inevitably culminating in the manifestation of human mental consciousness.

The Lagrangian governing the anthropic sector, where temperatures and electromagnetic interactions are conducive to complexity, takes the form of the heterotic string (Gr. *heterosis*, "hybrid vigor"), one of the five limits of the fundamental string in M-Theory; see Figure 15-1. The heterotic string is effectively a hard-wired connection between branes of differing bosonic and fermionic dimensionalities. Neither fundamental strings nor the braneworld is manifest in the eleven dimensional limit of the M-Theory symmetry space, where all dualities are resolved, but instead they emerge together through the breaking of this symmetry. The heterotic string underlies the maximally complex interplay of the particle species, their composites, and the gauge symmetry field interactions possible on the M-landscape.

The anthropic pathway follows the rise of complexity from the super quantum potential of the primordial universe, along the heterotic extremum of the M-landscape, to the highly complex organisms living within coevolved environmental niches on earth. The ability of an organism to sustain the low entropy required for complexification—notwithstanding the universal thermodynamic drift toward increasing entropy—arises as a product of cyclic interdependencies in the chemical reaction pathways of dissipative open systems.[297] Some sets of interdependent cycles are more efficient than others at consuming limited resources from the common chemical substrate. They are fitter for survival and population success in a competitive environment. The proficiency of cycle sets to consume resources, plus the ability to adapt to variations in the substrate, through the addition and tweaking of nested feedback/feedforward cycles, is the basis of natural selection.

The term natural selection was introduced in 1859 by Charles Darwin in his seminal work, *On the Origin of Species by Means of Natural Selection, or the Preservation of Favoured Races in the Struggle for Life.* It uncovered the foundation of evolutionary biology; see Chapters 3, 12 & 14. Natural selection is the gradual process whereby the traits (characteristics) of an organism become more or less prevalent in a population as a function of their effect on the fitness of the holistic organism in its interactions with the environment. Darwin's theory is applicable to those special sets of chemical reaction cycles that implement holistically beneficial traits. These traits are promoted through evolution in accordance with the natural selection algorithm, which represents the "strong form" of Darwinism. The strong form entails a three step process: 1) reproduction of a system with heritability of its traits, 2) some variation of the traits among offspring systems, and 3) selection of the most reproductively successful trait variants relative to environmental conditions.

> *If variations useful to any organic being do occur, assuredly individuals thus characterised will have the best chance of being preserved in the struggle for life; and from the strong principle of inheritance they will tend to produce offspring similarly characterised. This principle of preservation, I have called, for the sake of brevity, Natural Selection.*
>
> — Charles Darwin [298]

Darwin's term "natural selection," refers specifically to biological evolution. Natural selection has now been shown to be a subclass of a more general evolutionary process, called *universal Darwinism.* Collectively, universal Darwinism influences every step along the anthropic pathway. It is a basic driver of evolution, manifesting in such diverse applications as quantum

Darwinism, natural selection, neuronal group selection, social sciences, economics, epistemology, and cosmological natural selection.

> *Universal Darwinism considers a Darwinian process as a substrate-neutral algorithm that may be extracted from Darwin's theory of natural selection. This algorithm is composed of three steps: replication or copying, variations amongst the copies, and selective survival of the copies determined by which variations they possess. It has been shown that any entity embodying this algorithm will evolve and will evolve to be better at surviving.*
>
> — John Campbell [299]

The selection process emerges in the deepest levels of the quantum world, in a form called quantum Darwinism; see Chapter 3. At the outset of a cosmic cycle, the selection algorithm commences in the purely quantal substrate (the super quantum potential) of emerging wavefunction sub-wholes. Initially, sub-wholes transform collectively, as a coherent superposition of all possible ways charge and mass quanta conserved by the Lagrangian (a functional over the entire symmetry space) could manifest in the form of localized particles.

Under certain conditions, a specific solution to the Lagrangian's superposed ensemble is selected to manifest as a decohered particle (a sub-whole in the classical dimensionalities), venturing a first step on the anthropic pathway. In Chapter 17, it was noted that the decoherence-recoherence of a stable particle depends on the relationship between a sub-whole and the environment, which, in the deeper ontological interpretation, depends on active information coming from the phasal form of the quantum potential, repetitively instructing a sub-whole to draw energy from the vacuum.

The state into which a quantum superposition manifestly decoheres is termed a pointer state, referring metaphorically to the pointer on a dial-type, classical measuring instrument, but the term is applied generically to include measuring devices of any ilk. The pointer state that survives a quantum decoherence transition is a representation of the "weak form" of Darwinism, relative to biological natural selection, in that it conforms to the less stringent three step algorithm that is applicable to the quantum substrate: 1) the conserved properties, or characteristics, of a pointer state are passed on to offspring in future classical events, 2) the variation of the characteristics of the copied pointer states is with respect to their combinations, and 3) selection of successful variations is determined by wavefunction conjugation with the environment. While the algorithm of quantum Darwinism is weaker than that of natural selection—heritability of disseminated vs. conjoined characteristics, combinatorial vs. genetic offspring—the key principles of universal Darwinism are enforced. In both these provinces there is exclusive proliferation of the fittest adaptive systems relative to their particular environment.

Decoherence of a particle involves an interaction between a sub-whole and the whole. The probability for, or the fitness of, a coherent sub-whole to decohere as a particular particle solution to the holistic wave function is determined not only by the particle's potentiality to survive immersion in the environment, but also by its conjugate effect upon that environment. Under suitable conditions, the coherence-decoherence relationship between a particle and its environment establishes a reciprocative cycle, resulting in a stable particle continuously exchanging information about mass-charge properties with its environment.

The reciprocative exchange of information between a stable particle and its environment essentially establishes the particle as a processor of signs that are reflective of globally conserved quantum charges. The effect of a particle on the environment manifests through the emission of vibrational signs (bosons) into that environment, which is comprised of the condensate and the virtual particle vacuum fields of framed gauge theory, as well as other decohered fermionic particle cycles embossed upon those fields. Reciprocally, the particle absorbs vibrational signs emitted by other particles in quantal increments.

Exchange of bosonic signs among decohered fermionic particles opens the possibility for the formation of composite structures and emergence of intermingling levels of process-cycling and complexity. The complexity needed to be sustained within the delicate electrochemical process-structure of a living organism demands a tranquil environment, such as the biosphere of our planet. How did these hospitable environmental conditions arise? Let us chronicle the story of evolution again, this time traveling through the macrocosm along its extremum of complexity, the anthropic pathway.

Initially, composite structures and process pathways were precluded from sustaining stability because the gauge forces necessary to hold them together were overpowered by the thermal turbulence of the post-big bang environment. Cosmological cooling permitted big bang nucleosynthesis, fusing the nucleons (three quarks into a proton or neutron) and the lightest composite atomic nuclei (hydrogen, helium, and traces of lithium isotopes) into a rarefied, homogeneous gas. Under the force of gravity, the gas coalesced into giant metal-free clouds which subsequently condensed into population 3 stars (in cosmology, the term "metallicity" denotes all of the atomic elements excluding hydrogen and helium; and the term "population" specifies chronology of stellar generations, where indexed numbers increase from present to past).

Under the pull of their own internal gravitational forces, population 3 stars further condensed until nuclear chain-reactions triggered powerful supernova explosions (plus relativistic jets issuing from the back holes which were also condensing at the time), fusing and spewing elements 4 through 20, along with some H_2 molecules, into the metal-poor universe. Yet again, gravitation went to work, condensing the metal-poor stellar nebula into population 2 stars, which subsequently fused and exploded its contingent of elements into the cosmos, establishing conditions where electromagnetism was able to bind gravitationally coalescing elements into giant molecular clouds. But once again, the cycle repeated, with stellar nebula coalescing, condensing into stars, and eventually exploding as supernovae, each blast emitting as much light as a galaxy; while fusing together the remainder of the 92 stable elements, thereby completing the periodic table of our current metal-rich universe. Finally, metal-rich population 1 stars, including our own sun, consolidated under the compacting force of gravity.

Today, metal-rich pre-solar nebulae continue to gather and condense into population 1 stars. A population 1 star generally develops at the center of a spinning proto-planetary disc, possibly with a complement of orbiting planets constituting a solar system. Current statistical analyses indicate that quintillions of planets inhabit the universe, with 40 billion earth-sized planets orbiting population 1 stars in our galaxy alone.[300] The metals available to a planet in a solar system depend on its orbital distance from its central star. In our solar system, only Venus, Earth, and Mars contain the metals required to support life. The particular course of cosmological evolution, from the big bang, to metal-rich solar systems, to tranquil ecospheres, should be viewed from a perspective that fully appreciates the conditions that make living systems possible, for there are many other pathways through which evolution has ventured. For example, in the pulsar PSR B1509-58, the atomic elements are packed in a super-dense lattice immersed in a sea of electrons—hardly an environment where the delicate process-structures underlying life's complexity can arise.

After the full complement of atomic elements and some simple molecules which had bonded in the interstellar medium condensed to form the planet earth, gravity continued its work by stratifying these constituents into density-based layers. These layered earth spheres included the geosphere (lithosphere), the hydrosphere, the atmosphere, and the magnetosphere.

The earth spheres interacted along gradients empowered by gravitational, nuclear, and electromagnetic forces, thermodynamics, change of phase (solid, liquid, and gas), precipitation, and chemical deposition. Large-scale interdependencies among the earth spheres developed, endowing the macrocosmic environment of earth with complexity and a biosphere that encompassed self-organizing adaptive systems. Holistic closure of these complex adaptive systems followed, giving birth to the microcosmic worlds in which life and consciousness would now ascend.

* * *

Bibliography

References for Introduction

František Baluška and Stefano Mancuso; *Deep Evolutionary Origins of Neurobiology*; Communicative & Integrative Biology; Landes Bioscience; Taylor & Francis Group; Abingdon, UK; 2009

Evolution of the First Nervous Systems; The Journal of Experimental Biology; The Company of Biologists Ltd; Cambridge, UK; 2005.

Tomás J. Ryan and Seth G. N. Grant; *The Origin and Evolution of Synapses;* Nature Reviews I Neuroscience; Macmillan Publishers Ltd; New York, NY; 2009.

J. A. Westfall and J. C. Kinnamon; *A Second Sensory—Motor—Interneuron with Neurosecretory Granules in Hydra*; Journal of Neurocytology; Kluwer Academic Publishers; 1978.

Christophe Dupre and Rafael Yuste; *Non-overlapping Neural Networks in Hydra Vulgaris;* Neurotechnology Center, Department of Biological Sciences and Neuroscience, Columbia University; New York, NY; 2017.

Sarah M. Farris; *Evolution of Insect Mushroom Bodies: Old Clues, New Insights;* Arthropod Structure & Development; Elsevier Ltd.; 2005.

Dave K. Jacobs et al; *Evolution of Sensory Structures in Basal Metazoan*; Society for Integrative and Comparative Biology; Oxford University Press; UK; 2007.

Benjamin J. Liebeskind et al; *Evolution of Animal Neural Systems*; Annual Review of Ecology, Evolution, and Systematics; annualreviews.org; University of Texas; Austin, TX; 2017.

Todd E. Feinberg and Jon Mallatt; *The Evolutionary and Genetic Origins of Consciousness in the Cambrian Period over 500 Million Years Ago*; Frontiers in Psychology; Frontiers Media; Creative Commons Attribution; 2013.

Maria V. Sanchez-Vives et al; *Shaping the Default Activity Pattern of the Cortical Network* ; Neuron; Cell Press; Elsevier Inc.; 2017

Edmund T. Rolls; *An Attractor Network in the Hippocampus: Theory and Neurophysiology*; Learning & Memory; Cold Spring Harbor Laboratory Press; Cold Spring Harbor, NY; 2007.

Walter J. Freeman; *The Emergence of Mind and Emotion in the Evolution of Neocortex*; Department of Molecular & Cell Biology; University of California at Berkeley; Il Pensiero Scientifico Editore download; 2020.

Gerald M. Edelman and Joseph A. Gally; *Reentry: A Key Mechanism for Integration of Brain Function*; Frontiers in Integrative Neuroscience; Open Access; 2013.

Aaron Schurger et al; *Cortical Activity Is More Stable When Sensory Stimuli Are Consciously Perceived*; Proceedings of the National Academy of Sciences; Open Access; 2015.

Peter T. Walling; *An Update On Dimensions of Consciousness*; Proceedings, Baylor University Medical Center; Taylor & Francis Group; 2020.

Roger Orpwood; *Information and the Origin of Qualia*; Frontiers in systems neuroscience; Open Access; 2017.

John R. Searle; *Why I Am Not a Property Dualist; in Intentionality, an Essay in the Philosophy of Mind;* Cambridge University Press; Cambridge, MA; 1983.

Owen Flanagan; *Consciousness Reconsidered*; A Bradford Book; The MIT Press; Cambridge, MA; 1992.

* * *

References for Chapters 1-18

---

1 Jacob Schwichtenberg; *Physics from Symmetry*; Springer International Publishing; Cham, Switzerland; 2015.

2 Pierre Curie; *Sur la sym´etrie dans les ph´enom`enes physiques. Sym´etrie d'un champ ´electrique et d'un champ magn´etique*; Journal de Physique; 1894.

3 Ashoke Sen; *Tachyon Condensation on the Brane Antibrane System*; Mehta Research Institute of Mathematics and Mathematical Physics; Allahabad, India; 2008.

4 L. O' Raifeartaigh; *The Dawning of Gauge Theory*; Princeton University Press; Princeton, NJ; 1995.

5 Roger Penrose; *The Road to Reality*, pp., 726-732; Knopf, Borzoi Books, Random House, Inc.; New York, NY; 2004.

6 Ilya Prigogine and Isabelle Stengers; *Order out of Chaos: Man's New Dialog with Nature*; Bantam Books.; New York, NY; 1984.

7 Ilya Prigogine; *From Being to Becoming, Time And Complexity in the Physical Sciences;* W. H. Freeman and Company; London, UK; 1980.

8 Charles Darwin; *On the Origin of Species by Means of Natural Selection, or the Preservation of Favoured Races in the Struggle for Life*; John Murray; London, UK; 1859.

9 Francois Jacob; *Evolution and Tinkering;* Science, Volume 196; Washington, DC; 1977.

10 John Campbell; *Quantum Darwinism as a Darwinian Process;* Published online at Arxiv/General Physics/1001.0745; 2010.

11 Gerald Edelman; *Neural Darwinism, The Theory of Neuronal Group Selection;* Basic Books; New York, NY; 1987.

12 Wojciech Zurek; *Quantum Darwinism and Envariance*; From Barrow et al; *Science and Ultimate Reality: From Quantum to Cosmos*; Cambridge University Press; UK; 2004

13 M. Mitchell Waldrop; *Complexity: The Emerging Science at the Edge of Order and Chaos;* Simon and Schuster; New York; 1992.

14 William Martin and Michael J. Russell; *On the Origins of Cells: a Hypothesis for the Evolutionary Transitions from Abiotic Geochemistry to Chemoautotrophic Prokaryotes, and from Prokaryotes to Nucleated Cells;* Published online by The Royal Society; London, UK; 2002.

15 Stewart Kauffman; *At Home in the Universe*; Oxford University Press; UK; 1995.

16 Wim Hordijk, Jotun Hein, and Mike Steel; Figure 5-4 adapted from *Autocatalytic Sets and the Origin of Life*; University of Oxford; UK; 2010.

17 H. J. Morowitz; *Energy Flow in Biology*; Academic Press; New York, NY; 1968.

18 Vera Vasas et al; *Evolution Before Genes*; Biology Direct; BioMed Central; London, UK; 2012.

19 Nick Lane; *Life Ascending: The Ten Great Inventions of Evolution;* W. W. Norton and Co., Inc.; New York, NY; 2009.

20 Maria C. Rivera and James A. Lake; *The Ring of Life Provides Evidence for a Genome Fusion Origin of Eukaryotes;* Nature, Vol. 431; London, UK; 2004.

21 Lynn Margulis and Dorion Sagan; *Marvellous Microbes;* Resurgence 206: 10–12; Devon, UK; 2001.

22 Karl Niklas and Stuart Newman; *The Origins of Multicellular Organisms;* Evolution & Development, Vol. 15; Wiley-Blackwell; Hoboken, NJ; 2013.

23 Peter Coveney and Roger Highfield; *Frontiers of Complexity*; Ballantine Books, Random House; New York, NY; 1995.

24 Maddigan et al; *Biology of Microorganisms,* 13th ed.; Benjamin Cummings Publishing; San Francisco, CA; 2012.

25 John Tyler Bonner; *The Origins of Multicellularity;* Integrative Biology: Issues, News, and Reviews; Wiley-Liss, Inc.; Wilmington, DE; 1998.

26 Richard Grosberg and Richard Strathmann; *The Evolution of Multicellularity: A Minor Major Transition?*; Annual Reviews; Palo Alto, CA; 2007.

27 Gary G. Matthews; *Neurobiology: Molecules, Cells and Systems;* Blackwell Science, Inc.; Malden, MA; 2001.

28 Sakary et al; *A Post-Synaptic Scaffold at the Origin of the Animal Kingdom;* Creative Commons Attribution License; 2007.

29 Pawel Burkhardt and Simon Sprecher; *Evolutionary Origin of Synapses and Neurons – Bridging the Gap*; BioEssays; Wiley Periodicals, Inc.; Creative Commons Attribution License; 2017.

30 Leonid L. Moroz et al; *The Ctenophore Genome and the Evolutionary Origins of Neural Systems*; Nature, Vol. 510; 2014.

31 Robert O. Becker and Gary Seldon; *The Body Electric, Electromagnetism and the Foundation of Life*; Harper, William Morrow and Company, Inc.; New York, NY; 1985.

32 Thibaut Brunet and Detlev Arendt; *From Damage Response to Action Potentials: Early Evolution of Neural and Contractile Modules in Stem Eukaryotes;* Philosophical Transactions; Royal Society Publishing; London, UK; 2006.

33 Detlev Arendt, Harald Hausen, and Gunter Purschke; *The 'Division of Labour' Model of Eye Evolution*; Philosophical Transactions; Royal Society Publishing; London, UK; 2009.

34 Chen Chang-Po, Tseng Chin-Hsiu, and Bih-Yuh; *The Development of the Catecholaminergic Nervous System in Starfish and Sea Cucumber Larvae*; Chen Institute of Zoology; Taipei, Taiwan, R.O.C.; 1995.

35 Tomer, Denes, Tessmar-Raible, and Arendt; *Profiling by Image Registration Reveals Common Origin of Annelid Mushroom Bodies and Vertebrate Pallium;* Cell; Cambridge, MA; 2010.

36 Jon H. Kaas; *The Evolution of Complex Sensory Systems in Mammals*; Department of Psychology, Vanderbilt University; Journal of Experimental Biology; printed in Great Britain; The Company of Biologists Limited; 1989.

37 Randle W. Ware et al; *The Nerve Ring of the Nematode Caenorhabditis elegans: Sensory Input and Motor Output;* California Institute of Technology; Pasadena, California; 1975.

38 Gary G. Matthews; *Neurobiology*; Blackwell Science, Inc.; Malden, MA; 2001.

39 Google Images / Brain / labeled for reuse / 2016. Figures 6-1, 6-2, 6-3, and 6-5.

40 Grillner, Robertson, and Stephenson-Jones; *The Evolutionary Origin of the Vertebrate Basal Ganglia and its Role in Action Selection;* The Journal of Physiology; Wiley Online Library; 2013.

41 Suzana Herculano-Houzel; *The Human Brain in Numbers: a Linearly Scaled-up Primate Brain;* Frontiers in Human Neuroscience; www.frontiersin.org; 2009.

42 Paul D. MacLean; *The Triune Brain in Evolution: Role in Paleocerebral Functions;* Plenum Press; New York, NY; 1990.

43 Tanya M. Evans; *Brain Structural Integrity and Intrinsic Functional Connectivity Forecast 6 Year Longitudinal Growth in Children's Numerical Abilities;* The Journal of Neuroscience; Society for Neuroscience; Washington DC; 2015.

44 Gordon M. Shepard; Figure 6-4, adapted from *Neurobiology;* Oxford University Press, Inc.; New York, NY; 1983.

45 H. Supèr and H. B. M. Uylings; *The Early Differentiation of the Neocortex: a Hypothesis on Neocortical Evolution*; Oxford University Press; UK; 2001.

46 Chris C. King; *Cosmic Symmetry-breaking, Bifurcation, Fractality and Biogenesis;* Mathematics Department, University of Auckland; Auckland, New Zealand; 2004.

47 Erich Jantsch; *The Self-Organizing Universe;* Pergamon Press Ltd.; Oxford, UK; 1980.

48 Per Bak, Chao Tang and Kurt Weisenfeld; *Self-organized Criticality*; Physical Review; The Americal Physical Society; College Park, MD; 1988.

49 Gerhard Werner; *On Critical State Transitions between Different Levels in Neural Systems;* Department of Biomedical Engineering, University of Texas; Austin, TX; 2009.

50 György Buzsáki; *Rhythms of the Brain;* Oxford University Press, Inc.; New York, NY; 2006.

51 Lawrence M. Ward; Figure 6-6 adapted from *The Thalamic Dynamic Core Theory of Conscious Experience*; Consciousness and Cognition; Elsevier, B.V.; Amsterdam, Netherlands; 2011.

52 Francis Crick; Figure 6-7 adapted from *Function of the Thalamic Reticular Complex: The Searchlight Hypothesis;* Proceedings, National Academy of Sciences, Vol. 81, Neurobiology; Washington, DC; 1984.

53 Min Byoung-Kyong; *A Thalamic Reticular Networking Model of Consciousness*; Harvard Medical School; Boston, MA; 2010.

54 Figure 6-8 adapted from Francis Crick and Christof Koch; *Consciousness and Neuroscience* (Fuster's figure by Lippincott-Raven Publishers); The Salk Institute and California Institute of Technology; Pasadena, CA; 1998.

55 Olimpia Lombardi, et al; *What is Shannon Information?*; Philsci-archive. Pitt.edu; University of Pittsburgh; Pittsburgh, PA; open source; 2014.

56 Larry R. Squire and Stuart Zola-Morgan; *The Medial Temporal Lobe Memory System;* Science Magazine; New York, NY; 1991.

57 Larry R. Squire and Stuart Zola-Morgan; *Structure and Function of Declarative and Nondeclarative Memory Systems;* Proceedings of the National Academy of Sciences; Washington DC; 1996.

58 Alison R. Preston and Howard Eichenbaum; *Interplay of Hippocampus and Prefrontal Cortex in Memory;* Current Biology; Cell Press; Cambridge, MA.; 2013.

59 M. Marsel Mesulam; *From Sensation to Cognition*; Oxford University Press; UK; 1998.

60 Stanislas Dehaene; *Consciousness and the Brain;* Viking, Penguin Group; New York, NY; 2014.

61 Stephen Grossberg; *The Link between Brain Learning, Attention, and Consciousness;* Academic Press; Elsevier; Cambridge, MA; 1999.

62 C. Frith and R. J. Dolan; *Brain Mechanisms Associated with Top–down Processes in Perception*; Philosophical Transactions; Royal Society; London, UK; 1997.

63 Michael I. Posner and Marcus E. Raichle; *Images of Mind;* Scientific American Library, W. H. Freeman and Company; New York, NY; 1994.

64 Alexander Finglekurtz and Andrew Finglekurtz; *Brain-Mind Operational Architectonics Imaging: Technical and Methodological Aspects*; The Open Neuroimaging Journal; 2008.

65 E. O. Mann and O. Paulsen; *Role of GABAergic Inhibition in Hippocampal Network Oscillations;* Trends in Neuroscience, Cell Press; Elsevier; Cambridge, MA; 2007.

66 Andreas K. Engel et al; *Dynamic Predictions: Oscillations and Synchrony in Top-down Processing*; Nature Reviews – Neuroscience; Macmillan Magazines Ltd.; Holtzbrinck Publishing Group; Stuttgart, Germany; 2001.

67 Francisco Varela et al; *The Brainweb: Phase Synchronization and Large-scale integration;* Nature Reviews | Neuroscience; Macmillan Magazines Ltd.; London, UK; 2001.

68 Pascal Fries; *A Mechanism for Cognitive Dynamics: Neuronal Communication through Neuronal Coherence;* Trends in Cognitive Sciences; Elsevier; Amsterdam, Netherlands; 2005.

69 Timothy J. Buschman et al; *Synchronous Oscillatory Neural Ensembles for Rules in the Prefrontal Cortex;* Neuron Journal, Cell Press, Elsevier; Amsterdam, Netherlands; 2012.

70 Matthieu Ricard, Antoine Lutz, and Richard J. Davidson; *Mind of the Meditator;* Scientific American, Springer; New York, NY; 2014.

71 Gerald Edelman and Giulio Tononi; *A Universe of Consciousness. How Matter becomes Imagination;* Basic Books; New York, NY; 2000.

72 van Jelle Dijk and Gert-Jan Bleeker; Review of *A Universe of Consciousness. How Matter becomes Imagination;* Journal of Cognitive Systems Research, Elsevier; Amsterdam, Netherlands; 2001.

73 Jesper Hoffmeyer; *Biosemiotics;* University of Scranton Press; Chicago, IL; 2008.

74 Gerald Edelman; *The Remembered Present, A Biological Theory of Consciousness;* Basic Books; New York, NY; 1989.

75 Gerald Edelman; *Neural Darwinism, The Theory of Neuronal Group Selection;* Basic Books; New York, NY; 1987.

76 Giulio Tononi; *An Information Integration Theory of Consciousness;* Open access article, BioMed Central Ltd.; 2004.

77 George A. Mashour and Michael T. Alkire; *Evolution of Consciousness: Phylogeny, Ontogeny, and Emergence from General Anesthesia;* Proceedings of the National Academy of Sciences; Washington, DC; 2013.

78 Stanford Encyclopedia of Philosophy; *The Unity of Consciousness;* Stanford University; Stanford, CA; 2019

79 Tim Bayne; *The Unity of Consciousness;* Oxford University Press; UK; 2010.

80 S. Dehaen, M. Kerszberg, and J. P. Changeux; *A Neuronal Model of a Global Workspace in Effortful Cognitive Tasks*; Proceedings of the National Academy of Sciences of the United States of America; Neuroscience; Washington, D.C.; 1998.

81 Gonzalo Munevar; *Two Conscious Minds in the Same Brain? Split Brains and the Alien-Hand Syndrome*; Brain-mind magazine; Brain-mind Institute; Michigan State University; East Lansing, MI; 2012.

82 Erich Jantsch; *The Self-Organizing Universe;* Pergamon Press Ltd.; Oxford, UK; 1980.

83 Francis Crick and Christof Koch; *Consciousness and Neuroscience* (Fuster's figure by Lippincott-Raven Publishers); The Salk Institute and California Institute of Technology; Pasadena, CA; 1998.

84 D. Dennett; *Quining Qualia;* In A. Marcel and E. Bisiach, eds., Consciousness in Modern Science; Oxford University Press; UK; 1988.

85 Thomas Nagel; *What is it Like to be a Bat?*; The Philosophical Review; Duke University Press; Durham, NC; 1974.

86 Marek Marsel Mesulam; *From Sensation to Cognition;* Brain 121, 1013–1052; Oxford University Press; UK; 1998.

87 Lori Marinon et al; *Neuroanatomy of the Common Dolphin (Delphinus delphis) as Revealed by Magnetic Resonance Imaging (MRI);* The Anatomical Record 268: 411–429; Wiley-Liss, Inc.; 2002.

88 J. G. White et al; *The Structure of the Nervous System of the Nematode Caenorhabditis elegans;* Philosophical Transactions of the Royal Society; London; UK; 1986.

89 Randle W. Ware et al; *The Nerve Ring of the Nematode Caenorhabditis elegans: Sensory Input and Motor Output;* California Institute of Technology; Pasadena, California; 1975.

90 Andreas D. Peuke; *Correlations in concentrations, xylem and phloem flows, and partitioning of elements and ions in intact plants;* Journal of Experimental Botany, Vol. 61, No. 3, pp. 635–655; Oxford University Press; UK; 2010.

91 Huibo Ren et al; *Dynamic Analysis of ABA Accumulation in Relation to the rate of ABA Catabolism in Maize Tissues Under Water Deficit;* Journal of Experimental Botany, Vol. 58, No. 2, pp. 211–219; Oxford University Press; UK; 2007.

92 Sir John Woodroffe; *The Garland of Letters – Studies in the Mantra Shastra;* Ganesh & Company; Madras, India; 1974.

93 Swami Kuvalayananda; *Pranayama;* Science Press; Ephrata, PA; 1978.

94 György Buzsáki; *Rhythms of the Brain,* pp. 214-217; Oxford University Press, Inc.; New York, NY; 2006.

95 I. K. Taimni; *The Science of Yoga;* The Theosophical Publishing House; Madras, India; 1961.

96 W. Y. Evans-Wentz; *Tibetan Yoga and Secret Doctrines*; Oxford University Press; UK; 1958.

97 Jiddu Krishnamurti; 10th Public Talk; Bombay, India; 14th March, 1948.

98 William James; *The Principles of Psychology* (1890); Harvard University Press; Cambridge, MA; 1983.

99 Gerald Edelman; *Naturalizing Consciousness: A theoretical Framework;* Neuro-sciences Institute; San Diego, CA; 2003.

100 I. K. Taimni; *Man, God and the Universe;* The Theosophical Publishing House; Madras, India; 1969.

101 David J. Chalmers; *Facing Up to the Problem of Consciousness;* Research School of Social Sciences, Australian National University; Acton, Australia; 1995.

102 Arthur Peacocke; *The Sciences of Complexity: a New Theological Resource?*; Essay in *Information and the Nature of Reality*; Cambridge University Press; New York, NY; 2010.

103 Keith Ward; *God as the Ultimate Informational Principle*; essay in *Information and the Nature of Reality*; Cambridge University Press; New York, NY; 2010.

104 Edward Witten; *Magic, Mystery, and Matrix*; Notices of the AMS; 1998.

105 Rosalie Iemhoff; *Intuitionism in the Philosophy of Mathematics;* Stanford Encyclopedia of Philosophy; Stanford University; Stanford, CA; 2019.

106 Wesley Wrigley; *Gödelian Platonism and Mathematical Intuition*; European Journal of Philosophy; Wiley Online Library; 2021.

107 Michael Potter; *Was Gödel a Gödelian Platonist?*; Philosophia Mathematica; Oxford University Press; Oxford, UK; 2001.

108 Mark Colyvan; *The Indispensability of Mathematics*; Oxford University Press; Oxford, UK; 2001.

109 Anjan Chakravartty; *Scientific Realism*; Stanford Encyclopedia of Philosophy; Stanford University; Stanford, CA; 2017.

110 James Yorke; *Fifty Years of Chaos*; The Latecomer Institute—Longtermist Magazine; Open Philanthropy; San Francisco, CA; 2023.

111 Gregory J. Chaitin; *Gödel's Theorem and Information*; International Journal of Theoretical Physics; Springer Science + Business Media; Berlin/Heidelberg, GER; 1982.

112 Andrew Marks; *Computability theory*; Berkeley Department of Mathematics; University of California, Berkeley; Berkeley CA; 2004.

113 Mario Gomez-Torrente; *The Indefinability of Truth in the Wahrheit-sbegriff*; Annals of Pure and Applied Logic; 2004.

114 Marco Ragni, Gabriele Kern-Isberner, et al;; *Cognitive Logics – Features, Formalisms, and Challenges*; 24th European Conference on Artificial Intelligence; Santiago de Compostela, Spain; 2020.

115 P. C. Wason and T. Evans; *Dual processes in reasoning?*; Cognition; John Wiley & Sons, Inc.; Hoboken, NJ; 1974.

116 Jeff Buechner; *Are the Gödel incompleteness theorems limitative results for the neurosciences?*; Journal of Biological Physics; Springer Nature; Berlin/Heidelberg, GER; 2010.

117 Steven Hawking; *God Created The Integers: The Mathematical Breakthroughs that Changed History*; Penguin Books UK; 2006.

118 Ben Chaiken; *Gödel's Completeness and Incompleteness Theorems*; Department of Mathematics; University of Chicago; Chicago, IL; 2009.

---

119 Stephen Cole Kleen; *Introduction to Metamathematics;* Bibliotheca Mathematica; Wolters-Noordhoff Publishing and North-Holland Pub-lishing Com-pany; Oxford, UK and Amsterdam, NL; 1971.

120 Douglas R. Hofstadter; *Gödel, Escher, Bach: An Eternal Golden Braid;* Random House, Inc.; New York; 1979.

121 Robert P. Munafo; *Gödel's first Incompleteness Proof;* Creative Commons Attribution-NonCommercial 4.0 International License; https://mrob.com/cc-license.html; 2022.

122 Martin Hirzel; *On Formally Undecidable Propositions of Principia Mathematica and Related Systems;* https://hirzels.com/martin/papers/canon00-goedel.pdf; IBM Research; 2000.

123 Richard Kaye; *Arithmetic and the Incompleteness Theorems;* School of Mathematics and Statistics; The University of Birmingham; Birmingham, UK; 2000.

124 A. W. Moore; *Gödel's Theorem: A Very Short Introduction;* Oxford University Press; Oxford, UK; 2022.

125 Jason W. Steinmetz; *An Intuitively Complete Analysis of Gödel's Incompleteness;* Arxiv>math>https://arxiv.org/abs/1512.03667; Cornell University; 2020.

126 Martin Davis; *The Incompleteness Theorem;* Notices of the American Mathematical Society, Volume 53, Number 4; Providence, Rhode Island; 2006.

127 Ernest Nagel and James R. Newman; *Gödel's Proof;* New York University Press; New York; 2001.

128 Richard Tieszen; *Simply Gödel* (pp 59); Simply Charly; New York; 2017.

129 Solomon Feferman; *The nature and significance of Gödel's incompleteness theorems*; Institute for Advanced Study, Princeton; https://math.stanford.edu/~feferman/papers/Godel-IAS.pdf; 2006.

130 Christof Kreitz; *Unsolvable Problems in Logic*; Applied Logic Lecture 23: CS 4860; University of Potsdam; Potsdam, GER; 2003.

131 Serafim Batzoglou; *Gödel's Incompleteness Theorem*; arXiv:2112.06641v1 [math.HO]; 2021.

132 Albert A. Johnstone; *Self-reference and Gödel's Theorem: A Husserlian Analysis*; Published in *Husserl Studies* 19, pp. 131-151; Springer Publishing Company; New York; 2003.

133 Robert P. Munafo; *Gödel's First Incompleteness Proof at MROB*; this work is licensed under a Creative Commons Attribution-Non-Commercial 4.0 International License; 1996-2022.

134 W.V. Quine; *The Ways of Paradox and Other Essays*; Harvard University Press; Cambridge, Massachusetts; 1976.

135 Saul A. Kripke; *Gödel's Theorem and Direct Self-Reference*; arXiv:2010.11979v2 [math.LO]; 2021.

136 Peter Smith; *An Introduction to Gödel's Theorems*; Cambridge University Press; Cambridge, UK; 2007.

137 Peter Smith; *Gödel Without (Too Many) Tears*; Logic Matters; https://www.logicmatters.net/; Cambridge; 2022.

138 B. Meltzer and R. B. Braithwaite; *Kurt Gödel: On Formally Undecidable Propositions of Principia Mathematica and Related Systems 1*; Dover Publications; Garden City, NY; 1992.

139 Andrey Bovykin and Richard Kaye; *Order-types of models of Peano arithmetic: a short survey*; School of Mathematics and Statistics, University of Birmingham; Birmingham, UK; 2001.

140 Warren S. McCulloch and Walter Pitts; *A Logical Calculus of the Ideas Immanent in Nervous Activity*; Bulletin of Mathematical Biophysics; 1943.

141 Alan Turing; *Computing Machinery and Intelligence*; Mind; Oxford University Press; GB; 1950.

142 David Rumelhart, Geoffrey Hinton, and Ronald Williams; *Learning representations by back-propagating errors*; Nature Scientific Journal, volume 323, pages 533-536; London, UK; 1986.

143 John Rupert Firth; *Studies in Linguistic Analysis*; Linguistic Society of America; New York; 1961.

144 Paul Kirby; *A Brief History of AI*; The Kirby Foundation; AI Technology; https://thefusioneer.com; 2022.

145 Muhammad Azhar Iqbal et al; *Enabling the Internet of Things: Fundamentals Design, and Applications*; John Wiley & Sons Ltd.; Hoboken, NJ; 2021.

146 Rina Caballar and Cole Stryker; *What is neuromorphic computing?*; International Business Machines Corporation; Armonk, New York; https://www.ibm.com/think/topics/neuromorphic-computing; 2024.

147 Caxton Griffith Kibebe, Yue Liu, and Jiaxi Tang; *Harnessing optical advantages in computing: a review of current and future trends*; Frontiers Media SA; Lausanne, Switzerland; Creative Commons Attribution License; 2024.

148 Raveesh Gupta; *Quantum Computing and AI*; International Journal for Multi-disciplinary Research; Gujarat, India; 2024.

149 Anand Pagnis, Vedansh Tembhre, and Pranjal Vyas; *Technological Singularity – An Overview of What the Future Holds*; International Journal of Computer Engineering in Research Trends; IJCERT Publication House; open access journal; https://ijcert.org/ems/ijcert_papers/V8I1001.pdf; 2021.

150 James A. Reggia, Garrett E. Katz, and Gregory P. Davis; *Artificial Conscious Intelligence*; Journal of Artificial Intelligence and Consciousness; World Scientific Publishing Company, Marston Book Services; UK; 2020.

151 Arthur Juliani et al; *On the link between conscious function and general intelligence in humans and machines*; Journal of Machine Learning Research; JMLR, Inc. and Microtome Publishing; Brookline, MA; 2022.

152 Adam Safron; *Integrated World Modeling Theory (IWMT) Expanded: Implications for Theories of Consciousness and Artificial Intelligence*; The Society for the Improvement of Psychological Science, psyarxiv; 2020.

153 Rina Diane Caballar and Cole Stryker; *What is reasoning in AI?*; International Business Machines Corporation; Armonk, New York; https://www.ibm.com/think; 2026.

154 Erkam Guresen and Gulgun Kayakutlub; *Definition of artificial neural networks with comparison to other networks*; Elsevier Ltd.; Amsterdam, Netherlands; Academic Press, Pergamon Press; 2010.

155 José Mira; *Symbols versus connections: 50 years of artificial intelligence*; Neuro-computing, Volume 71, Issues 4–6, Pages 671-680; Science Direct, Elsevier; Amsterdam, Netherlands; 2008.

156 Din Asotic; *Neural Networks Simplified, Easy Introduction to Machine Learning and AI*; 2023.

157 Paul Rodrigues, Janet Wiles, and Jeffery L. Elman; *A Recurrent Neural Network that Learns to Count*; Connection Science; Open Access Journal; Taylor & Francis imprints; 1999.

158 Felix A. Gers, J®urgen Schmidhuber, Fred Cummins; *Learning to Forget: Continual Prediction with LSTM*; Proc. ICANN'99 Int. Conf. on Artificial Neural Networks (Edinburgh, Scotland), vol. 2, p. 850-855, IEE, London, UK; 1999.

159 Xia Zhao et al; *A review of convolutional neural networks in computer vision*; Artificial Intelligence Review; Springer Nature B.V.; Open Access, Creative Commons License; 2024.

160 Zhifei Zhang; *Derivation of Backpropagation in Convolutional Neural Network (CNN)*; University of Tennessee, Knoxville, TN; 2016.

161 Rukayat Sadiq; *CNN BackPropagation; Introduction to Deep Learning*; Carnegie Mellon University; Pittsburgh, PA; 2021.

162 Ashish Vaswani et al; *Attention Is All You Need*; 31st Conference on Neural Information Processing Systems; Long Beach, CA; 2017.

163 Christopher Bryant, Mariano Felice, and Ted Briscoe; *Automatic Annotation and Evaluation of Error Types for Grammatical Error Correction*; University of Cambridge; Cambridge, UK; 2017.

164 Dzmitry Bahdanau, KyungHyun Cho; Yoshua Bengio; *Neural Machine Translation by Jointly Learning to Align and Translate*; 3rd International Conference on Learning Representations; San Diego, CA; arXiv:1409.0473v7; 2015.

165 Roger Penrose; *The Emperor's New Mind*; Penguin Books; New York; 1991.

166 Rufn VanRullen and Ryota Kanai; *Deep Learning and the Global Workspace Theory*; The Brain and Cognition Research Center; University of Toulouse; Toulouse, FR; arXiv:2012.10390v2 [cs.AI]; 2021.

167 George A. Mashour et al; *Conscious Processing and the Global Neuronal Workspace Hypothesis*; Cell Press, Neuron; Cambridge MA;.2020.

168 Giulio Tononi, Melanie Boly, Marcello Massimini, and Christof Koch; *Integrated information theory: from consciousness to its physical substrate*; Perspectives, Nature Reviews I Neuroscience; Macmillan Publishers Limited; 2016.

169 Adam Safron; *Integrated World Modeling Theory*; Frontiers in Artificial Intelligence; Frontiers Media; Lausanne, Switzerland; 2020.

170 Steven Horst; *The Computational Theory of Mind*; Stanford Encyclopedia of Philosophy; Center for the Study of Language and Information; Stanford University; Stanford, CA; 2009.

171 Jaan Aru, Mototaka Suzuki, and Matthew E. Larkum; *Cellular Mechanisms of Conscious Processing*; Trends in Cognitive Sciences; Cell Press Reviews; Cambridge, MA; 2020.

172 Kenneth O. Stanley; *Neuroevolution: A different kind of deep learning*; O'Reilly Media, Inc.; 2017.

173 Risto Miikkulainen; *Neuroevolution*; Neural Network Research Group; Department of Computer Sciences, The University of Texas at Austin; 2022.

174 Shifei Ding et al; *Evolutionary artificial neural networks: a review*; Artificial Intelligence Review; Springer Science + Business Media B.V; 2011.

175 Julian Togelius; *Artificial General Intelligence*; The MIT Press; Cambridge, MA; 2024.

176 Karl Mason; *Advances in Evolutionary Neural Networks with Applications in Energy Systems and the Environment*; PhD Thesis; National University of Ireland, College of Engineering and Informatics; Galway, Ireland; 2018.

177 Justin K. Pugh, Lisa B. Soros, and Kenneth O. Stanley; *Quality Diversity: a new Frontier for evolutionary computation*; Frontiers in Robotics and AI; Open Access; Frontiers Media Limited; London, UK; 2016.

178 Kenneth Owen Stanley; *Efficient Evolution of Neural Networks through Complexification*; University of Texas; Austin, TX; 2004.

179 Sebastian Herzog, Christian Tetzlaff, and Florentin Wörgötter; *Evolving artificial neural networks with feedback*; Neural Networks; Science Direct; Elsevier; Amsterdam, NL; 2020.

180 Saman Khamesian and Hamed Malek; *Hybrid self-attention NEAT: a novel evolutionary self-attention approach to improve the NEAT algorithm in high dimensional inputs*; Springer Nature Link; [https://arxiv.org/pdf/2112.03670]; 2023.

181 Leon O. Chua; *Memristor-The missing circuit element*; IEEE Transactions on Circuit Theory; Institute of Electrical and Electronics Engineers; New York; 1971.

182 Fernando Aguirre et al; *Hardware Implementation of Memristor-based Artificial Neural Networks*; Nature Research; Springer Nature; open access; London, UK; 2024.

183 Nick Barney; *What is neuromorphic computing?*; Informa TechTaget; Newton, MA; 2024.

184 Nitin Rathi, Indranil Chakraborty, and Adarsh Kosta; *Exploring Neuromorphic Computing Based on Spiking Neural Networks: Algorithms to Hardware;*

185 Aditya Datar and Pramit Saha; *The Promise of Analog Deep Learning: Recent Advances, Challenges and Opportunities;* Institute of Electrical and Electronics Engineers; open access; New York; 2024.

186 Karim El-Laithy and Martin Bogdan; *Synchrony State Generation in Artificial Neural Networks with Stochastic Synapses;* Springer Nature; Berlin, GE; 2009.

187 Fangxin Liu et al; *SSTDP: Supervised Spike Timing Dependent Plasticity for Efficient Spiking Neural Network Training*; Frontiers in Neuroscience; Frontiers Media; Lausanne, Switzerland; 2021.

188 Francisco J. Romero et al; *Memcapacitor and Meminductor Circuit Emulators: A Review*; Electronics; Multidisciplinary Digital Publishing Institute; Basel, Switzerland; Open Access Journal; 2021.

189 Geert Van Damme; *Building an Object Oriented Neural Network*; Medium; medium.com, online publication; 2024.

190 Simona Ginsburg and Eva Jablonka; *The Transition to Experiencing: II. The Evolution of Associative Learning Based on Feelings*; Springer Science and Business Media LLC; Biological Theory; Berlin, Germany; 2007.

191 Tilman Rauker et al; *Toward Transparent AI: A Survey on Interpreting the Inner Structures of Deep Neural Networks;* https://arxiv.org/pdf/2207.13243; Cornell University; Ithaca, New York; 2023.

192 Vikas Hassija et al; *Interpreting Black-Box Models: A Review on Explainable Artificial Intelligence*; Springer Nature; Cognitive Computation; London, UK; open access; 2023.

193 Jonathan Birch, Alexandra K. Schnell, and Nicola S. Clayton; *Dimensions of Animal Consciousness;* Trends in Cognitive Sciences; Cell Press Reviews; Cambridge, MA; 2020.

194 Harvey B. Sarnat, Martin G. Netsky; *When does a ganglion become a brain? Evolutionary origin of the central nervous system*; Seminars in Pediatric Neurology, Volume 9, Issue 4; Science Direct; Elsevier; Amsterdam, NL; 2002.

195 Diego Becerra1, Andrea Calixto1, and Patricio Orio; *The Conscious Nematode: Exploring Hallmarks of Minimal Phenomenal Consciousness in Caenorhabditis Elegans;* International Journal of Psychological Research; Medellín, Colombia; 2023.

196 Eric Cassell; *"Directed Evolution": The Tiniest Brain Is Not Simple;* Evolution News and Science Today; Discovery Institute's Center for Science & Culture; Seattle, WA; 2025.

197 Dr. Wim Hordijk; *Autocatalytic Sets: Complexity at the Interface of Chemistry and Biology;* John Templeton Foundation; West Conshohocken, PA; 2022.

198 Wim Hordijk, Jotun Hein, and Mike Steel; *Autocatalytic Sets and the Origin of Life;* Entropy; Multidisciplinary Digital Publishing Institute; Open Access Journal; Basel, Switzerland; 2010.

199 Rotem Edri, Loren Dean Williams, and Moran Frenkel-Pinter; *From Catalysis of Evolution to Evolution of Catalysis*; Accounts of Chemical Research, Prebiotic Catalysis; American Chemical Society; Washington, DC; 2025.

200 Adam Safron; *Integrated World Modeling Theory (IWMT) Expanded: Implications for Theories of Consciousness and Artificial Intelligence*; The Society for the Improvement of Psychological Science; 2020.

201 Ramón Guevara, Diego M. Mateos, and José Luis Pérez Velázquez; *Consciousness as an Emergent Phenomenon: A Tale of Different Levels of Description*; Entropy; Multidisciplinary Digital Publishing Institute; open access; Basel, Switzerland; 2020.

202 Asmaa Rassil, Hiba Chougrad, and Hamid Zouaki; *Holistic Graph Neural Networks Based on a Global-based Attention Mechanism*; Knowledge-Based Systems, Volume 240; Science Direct, Elsevier; Amsterdam, Netherlands; 2022.

203 B Sanchez-Lengeling, E Reif, A Pearce, and AB Wiltschko; *A Gentle Introduction to Graph Neural Networks*; Distill publications; Creative Commons Attribution-licensed journals (Google Research); 2021.

204 Liane Gabora and Mike Steel; *A model of the transition to behavioral and cognitive modernity using reflexively autocatalytic networks*; Journal of The Royal Society Interface; London, UK; 2020.

205 Tamas Madl, Bernard J. Baars, Stan Franklin; *The Timing of the Cognitive Cycle*; open-access article; Creative Commons Attribution License; 2011.

206 Liane M. Gabora; *Conceptual Closure: How Memories are Woven into an Interconnected Worldview*; Annals of the New York Academy of Sciences; Wiley-Blackwell; Hoboken, NJ; 2000.

207 Shaojie Deng; *On the origin of biological teleonomy*; ScienceDirect, BioSystems; Elsevier; Amsterdam, Netherlands; 2026.

208 Osame Kinouchi and Mauro Copelli; *Optimal Dynamical Range of Excitable Networks at Criticality*; arXiv:q-bio/0601037v2; 2007.

209 Dietmar Plenz et al; *Self-Organized Criticality in the Brain*; Frontiers in Physics; Frontiers Media SA; Lausanne, Switzerland; open access; 2021.

210 Nike Walter* and Thilo Hinterberger; *Self-organized criticality as a framework for consciousness: A review study*; Frontiers in Psychology; Frontiers Media SA; Lausanne, Switzerland; open access; 2022.

211 Sagnik Sarkar et al; *Catalysis of neural activation functions: Adaptive feed-forward training for big data applications*; Springer Science+Business Media, LLC, part of Springer Nature; Berlin, Germany; 2022.

212 Jaan Aru, Matthew E. Larkum, and James M. Shine; *The feasibility of artificial consciousness through the lens of neuroscience*;
>quantitative biology>arXiv:2306.00915; Cornell University; Ithaca, New York; 2023.

213 Samson Abramsky et al; *Open Questions about Time and Self-reference in Living Systems*; https://arxiv.org/pdf/2508.11423v1; Cornell University; Ithaca, New York; 2025.

214 Andrea Roli, Johannes Jaeger, and Stuart A. Kauffman; *How Organisms Come to Know the World: Fundamental Limits on Artificial General Intelligence*; Frontiers in Ecology and Evolution; Frontiers Media SA; Lausanne, Switzerland; Creative Commons Attribution License; 2022.

215 Christopher Bryant, Mariano Felice, and Ted Briscoe; *Automatic Annotation and Evaluation of Error Types for Grammatical Error Correction*; University of Cambridge; Cambridge, UK.

216 AI Overview; *Systems-Based Reasoning*; Open Access; ScienceDirect; Elsevier; Amsterdam, Netherlands; 2025.

217 Serge Abiteboul; *Deductive and Object Oriented Databases*; National Institute for Research in Digital Science and Technology; Le Chesnay-Rocquencourt, FR; 1996.

218 Michael Kifer, George Lausen, and James Wu; *Logical Foundations of Object-Oriented and Frame-Based Languages*; Journal of the Association of Computing Machinery; New York; 1995.

219 Pamela Lyon et al; *Reframing cognition: getting down to biological-basics*; Philosophical Transactions of the Royal Society B; London, UK; 2021.

220 Simon van Gaal, Floris P. deLange, and Michael X Cohen; *The role of consciousness in cognitive control and decision making*; Frontiers Media; Frontiers in Human Neuroscience; Lausanne, Switzerland; 2012.

221 Jakob von Uexküll; Translated by Joseph D. O'Neil; *A Foray Into the Worlds of Animals and Humans: With A Theory of Meaning*; University of Minnesota Press; Minneapolis, MN; 2010.

222 Won Jeon; *Second-Order Recursions of First-Order Cybernetics: An "Experimental Epistemology"*; De Gruyter; Open Philosophy; Berlin, GE; 2022.

223 James Hoopes; *Peirce on Signs: Writings on Semiotic by Charles Sanders Peirce*; The University of North Carolina Press; Chapel Hill, NC; 1991.

224 Adam Safron; *The Radically Embodied Conscious Cybernetic Bayesian Brain: From Free Energy to Free Will and Back Again*; Entropy; Multidisciplinary Digital Publishing Institute; Open Access Journal; Basel, Switzerland; 2021.

225 Frank Jackson; *Epiphenomenal Qualia*; The Philosophical Quarterly; Oxford University Press, Wiley-Blackwell; Hoboken, NJ;1982.

226 Thomas Nagel; *What is it like to be a bat?*; The Philosophical Review; Duke University Press, Cornell University; Durham, NC; 1974.

227 Roger Orpwood; *Information and the Origin of Qualia*; Frontiers in Systems Neuroscience; Frontiers Media SA; Lausanne, Switzerland; Creative Commons Attribution License; 2017.

228 Elijah Chudnoff; *Cognitive Phenomenology (New Problems of Philosophy)*; Routledge, Taylor and Francis; New York; 2015.

229 Janet Levin; *Qualia*; Routledge Encyclopedia of Philosophy; Taylor and Francis; Oxfordshire, UK; 2011.

230 *Qualia*; Stanford Encyclopedia of Philosophy; Stanford University; Stanford, CA; Metaphysics Research Lab, CSLI; 2003.

231 Liane Gabora; *Autocatalytic closure in a cognitive system: A tentative scenario for the origin of culture*; Psycoloquy; American Psychological Association; Washington, DC; 1998.

232 Xerxes D. Arsiwalla; *Qualia and the Formal Structure of Meaning*; open access; Cornell University; arXiv:2405.01148v1 [q-bio.NC]; 2024.

233 Rob Grzywinski; *How to Think about Emergence*; LinkedIn; social networking service; Sunnyvale, CA; 2023.

234 Stephen Grossberg; *A Canonical Laminar Neocortical Circuit Whose Bottom-Up, Horizontal, and Top-Down Pathways Control Attention, Learning, and Prediction*; Frontiers in Systems Neuroscience; Frontiers Media SA; Lausanne, Switzerland; Creative Commons Attribution License; 2021.

235 Adam Frank; *Even the smartest AI likely won't be "alive." Here's why*; Big Think web portal; https://bigthink.com/13-8/is-life-comptuable/; 2024.

236 Robert Rosen; *Life Itself, A Comprehensive Inquiry into the Nature, Origin and Fabrication of Life*; Columbia University Press; New York; 1991.

237 Research publication; *Introduction to Anthropology: Holistic and Applied Research on Being Human*; Indiana University of Pennsylvania; Department of Anthropology; Indiana, PA; 2022.

238 Ernese Norelus; *Large Concept Models (LCMs) —Rethinking the Future of AI*; Medium.com; online publishing platform; 2025.

239 Liangwei Yang et al; *Cyclic Neural Network*; arXiv:2402.03332v1 [cs.NE]; 2024.

240 MathWorks; *Choosing the Best Machine Learning Classification Model and Avoiding Overfitting*; MathWorks Ebook, https://www.mathworks.com; Natick, MA; 2026.

241 Frank Gronwald and Friedrich W. Hehl; *On the Gauge Aspects of Gravity*; Institute for Theoretical Physics, University of Cologne; Koln, Germany; 1996.

242 Hong-Mo Chan and Tsou Sheung Tsun; *Exploring Framed Gauge Theory as Basis for Physical Models*; University of Oxford; UK; 2011.

243 Michael E. Peskin; *Supersymmetric Dark Matter in the Harsh Light of the Large Hadron Collider*; Stanford University, Menlo Park, CA; 2014.

244 S. Chatrchyan et al (CMS Collaboration, CERN); *Interpretation of Searches for Supersymmetry with Simplified Models*; Physical Review D; 2013.

245 Frank Gronwald; *Metric–Affine Gauge Theory of Gravity*; Institute for Theoretical Physics, University of Cologne; Koln, Germany; 1997.

246 Abhay Ashtekar; *Quantum Mechanics of Geometry*; Center for Gravitational Physics and Geometry, Pennsylvania State University; University Park, PA; 1999.

247 Shing-Tung Yau and Steve Nadis; *The Shape of Inner Space;* Basic Books, Perseus Group; New York, NY; 2010.

248 David Bohm and F. David Peat; *Science, Order, and Complexity;* Routledge Classics; Oxford, UK; 2011.

249 H. D. Zeh; *Quantum Nonlocality vs. Einstein Locality;* Web Essay; 2011.

250 Interpretations cited in *Holism and Nonseparability in Physics;* Stanford Encyclopedia of Philosophy; Center for the Study of Language and Information, Stanford University; Stanford, CA; 2016.

251 Juan Maldacena and Leonard Susskind; *Cool Horizons for Entangled Black Holes;* Institute for Advanced Study, Princeton, NJ, and Stanford Institute for Theoretical Physics, CA, resp.; 2013.

252 R. G. Chambers; *Shift of an Electron Interference Pattern by Enclosed Magnetic Flux;* Physical Review Letters; American Physical Society; College Park, MD; 1960.

253 Lynden K. Shalm et al; *Strong Loophole-Free Test of Local Realism;* Physical Review Letters; American Physical Society; College Park, MD; 2015.

254 B. J. Hiley; *The Early History of the Aharonov-Bohm Effect;* University of London; UK; 2013.

255 Steven Weinberg; *The First Three Minutes;* Basic Books; New York, NY; 1993.

256 Becker, Becker, and Schwarz; *String Theory and M-Theory,* Cambridge University Press; UK; 2007.

257 Joy Christian; *Why the Quantum Must Yield to Gravity,* Wolfson College, University of Oxford; UK; 1999.

258 Roger Penrose; *Cycles of Time;* Random House; London, UK; 2010.

259 *Webster's Third New International Dictionary of the English Language;* Merriam-Webster, Incorporated; Springfield, MA; 1993.

260 *Random House Unabridged Webster's Dictionary;* Collexion Reference Software; Lerneout & Hausple; 1998.

261 Rudolf Clausius; *The Mechanical Theory of Heat with its Applications to the Steam Engine and to the Physical Properties of Bodies*; John Van Voorst; London, UK; 167.

262 E. Michaelides; *Entropy, Order and Disorder;* The Open Thermodynamics Journal, Bentham Science Publishers Ltd; 2008.

263 Leonard Susskind; *The Black Hole War*; Back Bay Books; Little, Brown and Company; New York, NY; 2008.

264 Paul J. Steinhardt and Neil Turok; *Endless Universe;* Doubleday, Random House, Inc.; New York, NY; 2007.

265 Jean-Luc Lehners; *Ekpyrotic and Cyclic Cosmology,* Princeton Center for Theoretical Science; Princeton, NJ; 2009.

266 Brian Green; *The Elegant Universe;* Random House; New York, NY; 1999.

267 Fernando Quevedo; *Lectures on String/Brane Cosmology;* Centre for Mathematical Sciences; University of Cambridge; UK; 2002.

268 Lisa Randall; *Warped Passages;* Harper Perennial; New York, NY; 2006.

269 Paul J. Steinhardt and Neil Turok; *Cosmic Evolution in a Cyclic Universe;* Princeton University; Princeton, NJ; 2002.

270 Ichiro Oda;; *Emergence of Superstring from Pure Spinor*; University of the Ryukyus; Okinawa, Japan; 2011.

271 Ashoke Sen; *Tachyon Condensation on the Brane Antibrane System*; Mehta Research Institute of Mathematics and Mathematical Physics; Allahabad, India; 2008.

272 Emmy Noether; *Invariante Variationsprobleme*; Nachrichten Konigliche Gesellschaft der Wissenschaften zu Gottingen; July 16, 1918.

273 Nina Byers; *E. Noether's Discovery of the Deep Connection between Symmetries and Conservation Laws*; Physics Department, UCLA; Los Angeles, CA; 1998.

274 Holger Lyre; *Holism and Structuralism in U(1) Gauge Theory*; University of Bonn; Germany; 2004.

275 Hermann Minkowski; address delivered at the 80th Assembly of German Natural Scientists and Physicians, September 21, 1908.

276 Stephen Hawking; *The Universe in a Nutshell*; Bantam Books; New York, NY; 2001.

277 Edmund Copeland, James Lidsey, and Shuntaro Mizuno; *Correspondence between Loop-inspired and Braneworld Cosmology;* University of Nottingham, UK, University of London , UK, and Waseda University, Tokyo, Japan, resp.; 2008.

278 Martin Bojowald; *Loop Quantum Cosmology;* Max Planck Institute for Gravitational Physics; Munich, Germany; 2005.

279 Roger Penrose; *Quantum Theory and Beyond;* Cambridge University Press; UK; 1971.

280 Abhay Ashtekar and Jerzy Lewandowski; *Relation between Polymer and Fock Excitations*; Pennsylvania State University, University Park, PA, and Max Planck Institut f¨ur Gravitationsphysik, Albert Einstein Institut, Golm, Germany, resp.; 2001.

281 Mae-Wan Ho; Figure 5-1 adapted from *Thinking Again of Life's Miracle;* Lecture at Schumacher College 10th Anniversary Celebration; The Prince's Trust, Charlotte Road, London, UK; 2001.

282 Mae-Wan Ho; *The Rainbow and the Worm, The Physics of Organisms*; World Scientific; London, UK; 2008.

283 J. D. Bekenstein; *Information in the Holographic Universe;* Scientific American, Inc.; New York, NY; 2003.

284 Ilya Prigogine and Isabelle Stengers; *Order out of Chaos: Man's New Dialog with Nature*; Bantam Books; New York, NY; 1984.

285 Peter Coveney and Roger Highfield; *Frontiers of Complexity*; Ballantine Books, Random House; New York, NY; 1995.

286 Erwin Schroedinger; *What is Life?*; Cambridge University Press; UK; 1945.

287 Stewart Kauffman; *At Home in the Universe*; Oxford University Press; UK; 1995.

288 S. Lifson; *On the Crucial Stages in the Origin of Animate Matter*; Journal of Molecular Evolution; Springer; New York, NY; 1997.

289 Richard Dawkins; *The Selfish Gene;* Oxford University Press; UK; 1979.

290 Lynn Margulis and Dorion Sagan; *Marvellous Microbes;* Resurgence 206: 10–12; Devon, UK; 2001.

291 David Bohm; *Wholeness and the Implicate Order*; Routledge; London, UK; 1980.

292 David Bohm and B. G. Hiley; *The Undivided Universe*; Routledge; London, UK; 1993.

293 David Bohm; *Hidden Variables and the Implicate Order,* in Hiley and Peat; *Quantum Implications – Essays in Honor of David Bohm*; Routledge & Kegan Paul, LTD; London, UK; 1987.

294 David Bohm; *Quantum Theory;* Prentice Hall, Inc.; Englewood Cliffs, NJ; 1951.

295 Werner Heisenberg; *Physics and Beyond: Encounters and Conversations;* World Perspectives vol. 42; Harper & Row; New York, NY; 1971.

296 Leonard Susskind; *The Cosmic Landscape*; Back Bay Books; New York, NY; 2006.

297 Ilya Prigogine; *From Being to Becoming, Time And Complexity in the Physical Sciences;* W. H. Freeman and Company; London, UK; 1980.

298 Charles Darwin; *On the Origin of Species by Means of Natural Selection, or the Preservation of Favoured Races in the Struggle for Life*; John Murray; London, UK; 1859.

299 John Campbell; *Quantum Darwinism as a Darwinian Process;* Published online at Arxiv/General Physics/1001.0745; 2010.

300 Ian Stewart; *Calculating the Cosmos, How Mathematics Unveils the Universe;* Basic Books; New York, NY; 2016.

Made in the USA
Las Vegas, NV
26 June 2026

49937560R00229